Carlson

THE GEOLOGY OF KAUAI AND NIIHAU

BY

NORMAN E. A. HINDS

BERNICE P. BISHOP MUSEUM

BULLETIN 71

HONOLULU, HAWAII
PUBLISHED BY THE MUSEUM
1930

KRAUS REPRINT
Millwood, N.Y.
1985

Library of Congress Cataloging in Publication Data

Hinds, Norman E. A.
The geology of Kauai and Niihau.

Reprint. Originally published: Honolulu, Hawaii : Bernice P. Bishop Museum, 1930. (Bernice P. Bishop Museum bulletin ; 71)
Bibliography: p.
1. Geology--Hawaii--Kauai. 2. Geology--Hawaii--Niihau. I. Title. II. Series: Bernice P. Bishop Museum bulletin ; 71)
QE349.H3H56 1985 559.69'4 85-74
ISBN 0-527-02177-6

Reprinted with the permission of the original publisher

Printed in the U.S.A.

CONTENTS

ILLUSTRATIONS

The Geology of Kauai and Niihau

By

Norman E. A. Hinds

INTRODUCTION

SCOPE AND PURPOSE

Geological investigations in Hawaii have been very largely centered on the volcanic sink at Kilauea. The opportunities for the study of volcanic mechanism in this great, natural laboratory have proved more attractive to most field workers than the rather prosaic tack of unravelling the history of the extinct eruptive centers of this group. Before the beginning of the present decade, relatively little intensive geological work had been done in other parts of the archipelago; many reconnaissance observations had been published, but no area had been thoroughly investigated. The most valuable general reports are those of Dana (19)[1], who visited Hawaii, Oahu, and Kauai in 1840 and Hawaii, Maui, and Oahu in 1887; and of Dutton (23), who made a brief reconnaissance of parts of Hawaii, Maui, Oahu, and Kauai in 1882. Though these investigators spent only brief periods in the field, their recorded observations will serve as the basis for all later geologic and morphologic studies in Hawaii and in similar volcanic regions. Their reports discuss the nature of volcanic action, and describe the general characters of the lavas, but petrographic descriptions are not included. In a later publication, Dana (21) summarized his view of the geology of the Hawaiian islands and gave the microscopic characters of a few lavas from the islands of Hawaii, Maui, and Oahu. Two papers by Hitchcock (38) are largely confined to a description of the more recently developed geologic and morphologic features of Oahu, especially in the vicinity of Honolulu. A work by Bryan (5) contains a chapter on the geology of the leeward and windward Hawaiian islands.

Though Hawaii is the best known and the most easily accessible volcanic group in the Pacific, and information regarding the various phases of its geology is needed as a basis for studies on other volcanic islands, no thoroughgoing petrographic descriptions of systematic rock collections from any of its principal eruptive centers have been made. The recent contributions by Washington (58) to the petrology of the archipelago are of great value, but these should be supplemented by miscroscopic examination and chemical analyses

[1] The numbers in parentheses refer to Bibliography on pages 102-103.

of more extensive suites which represent thorough samplings of the eruptive and irruptive rocks of the various domes. Only by such means can the geologist be sure that important rock types have not been overlooked and that discussions of the genesis of the lavas are based upon a complete knowledge of the varieties composing the various eruptive centers. As the area of the archipelago is little more than 6,000 square miles, intensive field studies carried on for a few years would yield the facts which are most desired at the present time. Fortunately the Bernice P. Bishop Museum in Honolulu is sponsoring a number of such investigations. Under its auspices, Wentworth (60, 61) has published bulletins on the geology of Lanai and of Oahu; Palmer (48) has examined a number of leeward islands; and Stone (57) has described the geology of Kilauea.

Some years ago, Professor R. A. Daly urged me to take part in a systematic study of Hawaiian geology by making a detailed examination of one or more of the islands. This plan was later made possible through a liberal grant by the Committee on Sheldon Fellowships of Harvard University. The island of Kauai was selected because it has been extensively dissected and therefore seemed to offer the best opportunity for deciphering the subsurface volcanic structures, for the collection of a representative suite of igneous and sedimentary rocks, and for observation regarding the geomorphic development of a lava dome which long since has become extinct. Also, the United States Geological Survey, in 1911, had published an excellent map of the island, thus supplying a topographic base for the geologic studies. In 1921 I spent six months in the field and in 1922 was able to continue the investigation during the three months of the summer as a Bishop Museum Fellow in Yale University. The greater part of the two seasons was spent on Kauai. In addition, I made a reconnaissance survey of Niihau, and brief comparative studies on Oahu, Maui, and Hawaii.

The chief problems studied in the field were:

1. The subsurface volcanic structures and the nature of the igneous mechanism.

2. The lavas and pyroclastic deposits erupted long after the close of the principal eruptive period and after most of the existing topographic features had been developed.

3. The topography as affected by age, types of erosion, and faulting. Because little attention has heretofore been paid to the development of the relief of extinct or quiescent lava domes, I considered this one of the major field problems.

4. The sedimentary rocks and the conditions of sedimentation.

Extensive suites of the igneous and sedimentary rocks, the weathered lavas and various types of soils, and such fossils as are present were collected for laboratory study.

The present report discusses the morphology and general geology of Kauai and, in less detail, of Niihau. In papers to be published later I propose to deal with the petrography of the igneous rocks, the volcanological problems, the nature of rock weathering, the formation of soils in Hawaii, and also to give a more complete description of the sediments.

ACKNOWLEDGMENTS

The field studies embodied in this report were undertaken at the suggestion of Professor R. A. Daly, and under his inspiration and unfailing guidance, the assembling of results has been carried to completion. Throughout the course of the work, I have been aided in many ways by others, and to them I offer hearty thanks. Professors J. E. Wolff, W. M. Davis, and R. de C. Ward have given valuable counsel on many of the problems. Professor H. E. Gregory, Director of the Bernice P. Bishop Museum, placed the facilities of that institution at my disposal and assisted in securing equipment for the field work. Dr. H. S. Washington is analyzing a number of the igneous rocks in connection with his studies of Hawaiian petrology. Professor Junius Henderson of the University of Colorado has identified the fossils collected on Kauai and Niihau. Mr. A. O. Burkland, in charge of the Hawaiian branch of the United States Geological Survey, rendered valuable assistance in many ways. Climatological data were furnished by Mr. L. H. Daingerfield, formerly in charge of the Hawaiian Section of the United States Weather Bureau.

While in the field, I found the people of Hawaii extremely cordial and ready to cooperate in carrying on the research whenever possible. I should have been unable to examine a very considerable part of Kauai had it not been for guides, horses, and lodging places furnished by the plantation authorities and by others. Special acknowledgment must be accorded to the late J. M. Lydgate, who constantly advised and guided me during my work on Kauai. Through the courtesy of Mr. Aubrey Robinson and family, a visit to Niihau was made possible, and also an examination of the Kaholuamanu region on Kauai. Thanks are also due Mrs. J. M. Lydgate, Mrs. D. R. Isenberg, Mr. F. J. Dollinger, Mr. R. D. Moler, Mr. W. H. Rice, Judge L. A. Dickey, Mr. L. D. Larsen, Mr. F. A. Alexander, Mr. B. D. Baldwin, Mr. H. P. Faye, and Mr. W. J. Hardy of Kauai; to Mr. J. Waterhouse, the late Mr. Alonzo Gartley, and Professor H. S. Palmer of Oahu; to Mr. H. A. Baldwin, Mr. B. Williams, Dr. G. S. Aikens, Mr. R. Penhallow, and Mr. L. Smith of Maui; and to Dr. T. A. Jagger, Jr., and Mr. R. H. Finch. To Mr. E. P. Lydgate, my assistant in the field during two seasons, thanks are due for loyal and capable service. I am indebted to Mr. R. D. Russell of the University of California for assistance and suggestions in the preparation of the manuscript.

PREVIOUS GEOLOGICAL INVESTIGATIONS

Kauai and Niihau lie off the main line of travel in Hawaii, and hence have not been visited by many geologists. Niihau is privately owned, and is closed to the public.

In 1840, Dana spent four days on Kauai and visited a number of localities. His report (21) gives the most complete and accurate account of the general geologic and morphologic features of the island; though, in the light of more extended investigations, certain of his conclusions must be revised. Dana (18) summarized his early sketch of the geology of Kauai in a volume published after his second visit to Hawaii in 1887. Dutton (23) spent a short time on Kauai in 1882, but makes only casual references to the island. Cross (7) in 1902 and Powers (52) in 1915 collected igneous rocks from Kauai, and their reports mention certain of the geological features. Two papers by Powers (52, 53) contain brief notes on the geology of Niihau. Bryan (5) and Hitchcock (40) describe the general characters of the geology of both Kauai and Niihau.

GEOGRAPHY OF KAUAI

GENERAL FEATURES

The Hawaiian Archipelago, consisting of two distinct sections, the windward and the leeward, extends for more than 1,900 miles along the northern margin of the tropical climatic belt. The windward islands lie between latitudes 18° 54′ and 22° 15′ north; the leeward islands between 23° and 28° 15′ north. The windward section is composed of five large and three small islands extending from Kauai and Niihau southeastward to Hawaii—a distance of about 300 miles (fig. 1). Excepting Kahoolawe, the windward islands are inhabited.

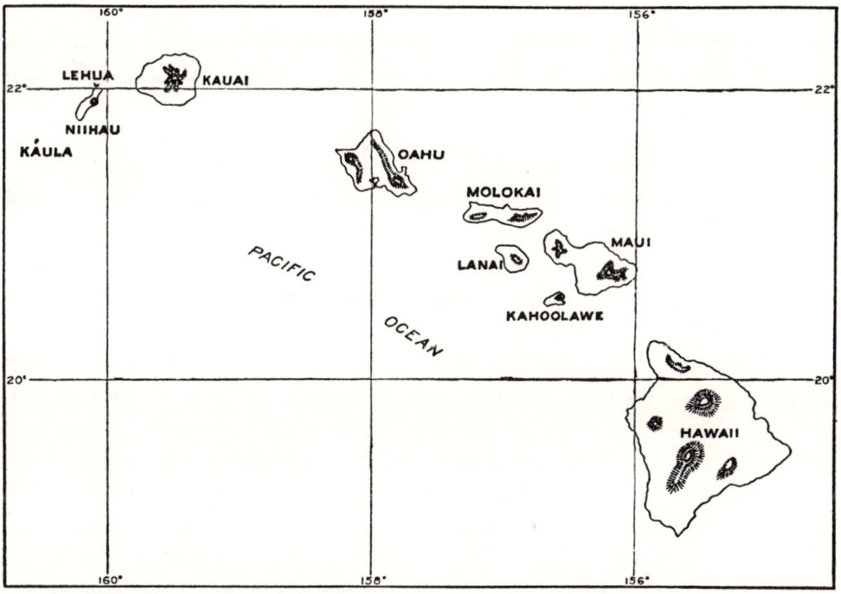

FIGURE 1.—The windward section of the Hawaiian Archipelago. From U. S. Coast and Geodetic Survey, Chart No. 4102.

West of Kauai and Oahu the archipelago is continued for more than 1,300 miles by 13 islets which have a total area of about 6 square miles. At the eastern end of this leeward group are four volcanic stacks with bordering reefs; the highest stack rises 895 feet above sea level. To the westward are atolls and calcareous sand islands with which no remnants of volcanic rocks are associated. In detail, the leeward group is composed of 6 bank atolls, 4 banks with cliffed central islands or stacks of lava partially surrounded by reefs, 5 banks without reefs or central islands, and 2 intermediate forms. The banks are 10 to 30 miles in diameter; the maximum depth of water over their surfaces in few places exceeds 30 or 40 fathoms. On Midway Island is an important American cable station; otherwise the leeward islands are uninhabited. Eight islands of the group have been set aside as a Federal Bird Reservation.

A deep channel, 68 miles in width, lies between Kauai and Oahu, and Nawiliwili, the nearest landing on Kauai is 90 miles from Honolulu. Semi-desert Niihau is a comparatively small island about 20 miles southwest of Kauai. On unusually clear days, the summits of the western mountains of Oahu are dimly visible from the eastern shore of Kauai, and, from the western side of Kauai, the cliffed dome remnant of Niihau usually can be seen.

Kauai is a deeply eroded, subcircular island, fourth in size in the Hawaiian Archipelago. Its area is 547 square miles. The maximum east-west diameter of Kauai is 32 miles; the maximum north-south diameter, 22 miles. The highest elevations, located near the center of the island, are about 5,000 feet. Adjoining Kauai is the island of Niihau with its associated tuff cones, Lehua and Kaula. (See p. 96.)

POPULATION

The chief elements in the population of Kauai, which was 29,247 according to the census of 1920, in order of numbers are Japanese, Chinese, Hawaiians, Portuguese, and Americans. As elsewhere in Hawaii, the pure native element is decreasing and that of mixed blood is increasing by intermarriage of the full-blooded Hawaiians with the whites and the Chinese. The number of Japanese is also being rapidly augmented because of the high birth rate among that people. Most of the population lives in small towns or plantation settlements, located on the lowlands of northern, eastern, and southern Kauai, or in scattered dwellings on the alluviated flats of the mouths of the river valleys crossing the lowlands. The chief towns are Kilauea, Kapaa, Lihue, Koloa, Hanapepe, and Waimea. Lihue, the county seat, is the principal town, though Waimea has a slightly larger population. The mountain section, because of its extreme ruggedness, inaccessibility, and in certain parts, because of its rainy climate, is almost uninhabited.

INDUSTRIES

The chief industries of Kauai are those common elsewhere in Hawaii—the cultivation of sugar cane, pineapples, and rice; fishing and cattle raising. The sugar plantations, which constitute the principal industrial enterprises, are confined to the lowlands except on the southern side of the island where fields extend to elevations of 1,100 feet. Apparently there is no physical reason why sugar cane should not be grown at higher elevations, as valuable crops are elsewhere raised at elevations of 2,000 feet. On Kauai however, the difficulties of cultivation, irrigation, and transportation over the rugged upland slopes are serious drawbacks. On the windward side, the greater amount of cloudiness and precipitation and the consequently somewhat lower average temperatures along the inner margins of the lowlands limit the elevation at which cane can be profitably grown. The crop requires long continued hot weather as well as large quantities of water. The planters find the soil and climate most favorable on the lower leeward (southern and southwestern) slopes, and there have developed the most successful plantations. On the windward lowlands, the yield of cane per acre is less; more intensive fertilization of the soil is required; replanting is necessary at frequent intervals; and every few years the fields must be allowed to lie fallow for a period of several months. Along the northern shore, where the lowlands are most heavily drenched with rain, conditions are least favorable for the growth of cane; in this section there is but one plantation as compared with six along the southern shore and three on the eastern side.

In the past, pineapples have been raised chiefly in localities where sugar cane could not be successfully grown. Since the World War, however, the acreage of cane fields has decreased and much of this land, as well as many formerly untilled areas, is now devoted to the cultivation of pineapples. The crop requires less attention and less irrigation then does sugar cane, and the demand for the canned product is growing. The pineapple industry bids fair to develop rapidly throughout Hawaii, though it will never replace the cultivation of sugar cane as the main industry. The main centers of pineapple growing on Kauai are near Lawai and in the Moloaa district. Canneries have been built at Lawai and Kapaa.

Rice culture is confined to small areas in the valley flats; the crop is not sufficient to supply the inhabitants and large quantities are imported. Some taro, from which the native food, poi, is made, is also grown in the valleys. Both taro and rice require great amounts of water, hence they can only be raised on the alluviated flats in the mouths of valleys where plenty of water is obtainable.

Fishing is actively carried on along the shores and in the adjacent deeper waters by the Japanese and Hawaiians. Several cattle ranches are located in various sections of the island, and considerable numbers of cattle and horses are raised.

Copra is produced in small quantities on eastern Kauai. Attempts to cultivate the coffee berry and sisal plant have not been successful on a commercial scale.

IRRIGATION

In spite of the heavy rainfall over most of Kauai, the sugar, rice, and taro crops require large additional quantities of water for their growth. Even before the coming of the white men, the natives constructed small ditch systems through their taro fields. At the present time, irrigation is practiced on a large scale, and ditch systems have been constructed in nearly all valleys between the Waimea and Wainiha rivers to carry waters from the upper reaches of the streams to the cane or rice fields on the lowlands or in the valley bottoms. The various ditch systems have been described by Martin and Pierce (45). A number of reservoirs have been built to store waters for irrigation and power; most of these are on the southern lowland in the Koloa and Hanapepe districts. The reservoir of the Koloa Sugar Company has a capacity of two and a half billion gallons—the largest on the island. The McBryde Sugar Company maintains pumping plants in Hanapepe and Lawai valleys, operated by a power station in Wainiha Canyon on the northern side of the island. The wells in Hanapepe Valley are about 3 miles from the sea and are more than 100 feet deep. In this system are three principal shafts, and more than 1,000 feet of tunnel. They supply about 20,000,000 gallons of water daily to the cane fields in these valleys. The same company drilled wells nearer the sea coast, but abandoned them because the waters in a short time became saline. The pumping plant of the Koloa Sugar Company which will raise 3,000,000 gallons every 24 hours is also operated by electricity.

Many wells have been drilled at various places on the lowlands for irrigation and domestic purposes. In those sunk near to or below sea level, the water has become brackish or saline in a relatively short time, indicating an active penetration of the lavas by sea water which is retarded only when the subsurface percolation of fresh water is not impeded. These wells are useless after the original supply of fresh water is drawn off.

HARBORS AND ROADS

The principal embayments in the Kauai coast are at the mouths of the Hanalei, Anahola, Hanamaulu, Huleia, Hanapepe, and Waimea rivers. No indentations of consequence are present along the cliffed northwestern coast or in the plain of western and southwestern Kauai. The chief stopping place for interisland ships is Ahukini in Hanamaulu Bay, the only place on the island where steamers can make fast to a wharf. From ships anchored off shore, passengers and freight may be landed in small boats at Waimea, Makaweli, Eleele, Koloa, and Port Allen on the south side of the island; at Kealia and Kilauea on the east coast; and at Hanalei. Work in progress at Nawiliwili Bay is designed to provide for Kauai a harbor which will permit large ships to anchor in safety at docks easily accessible from the back country.

A good road, paved most of the way, has been constructed along the coast nearly around the island from Barking Sands to Haena. Elsewhere the roads generally are not paved, and are mainly confined to the lowlands.

A road extends from Waimea to Kokee in the north central highlands where a number of mountain houses have been built, and roads have been built for some distance up most of the larger canyons. The dirt roads generally are in fair condition during the dry season, but frequently become impassable in the rainy months. Trails are abundant where the vegetation is scanty; in the jungle section the traveler generally must cut his way.

CLIMATE

Regional Features

The Hawaiian islands are geologic and geomorphic units lying well within the belt of the northeast trades. Over their surfaces, the winds blow with a more or less constant velocity; moderate to extreme differences of temperature from sea level to the summits of the domes prevail depending upon their elevations; and in most places there are extraordinary variations in the annual rainfall within short distances over the same mountain.

It is difficult to find anywhere greater constancy of atmospheric conditions than those prevailing over the ocean in the heart of the Trade Wind belt, where the Hawaiian Archipelago is located. The temperature varies little from summer to winter, and the general drift of the atmosphere is rarely interrupted by storms of great violence. Yet over the islands themselves "it may be said that there are almost as many climates as there are square leagues"; striking climatic contrasts are to be found in areas only a few miles apart.

According to Daingerfield (8) these local differences result from wide variations in rainfall over different sections of the islands and from the greater variability of the temperatures at the higher elevations. It is possible to quickly pass from the persistent summer of the sea coast and the lower mountain slopes to the chill climates of the highest summits, from highly arid regions to areas of frequent, heavy rains, from dominant sunshine to almost continuously cloud-draped mountain crests. The rainfall, which is the most variable of the climatic elements, is heaviest over the windward slopes; over the leeward portions of the various domes the precipitation is relatively light. The generally prevalent northeast winds are occasionally displaced for periods of a few days by southerly or southwesterly (Kona) winds which often blow with great violence and are frequently accompanied by unusually heavy rains over the leeward sections of the domes. Henry (31) considers the Kona storms to be the troughs of cyclonic depressions whose centers move eastward to the north of Hawaii. They travel slowly, and may require several days for their passage through the group. Though Kona storms are infrequent and irregular in their occurrence, they greatly affect the precipitation; the rainfall during a single one of these storms may exceed the annual average for a given section. Thus very striking contrasts are afforded in the records of annual precipitation at various stations during a series of years.

As pointed out by Daingerfield (8), the temperatures depend almost wholly on elevation, though they are affected

> . . . in a minor way by the slope and exposure to the wind. The range of mean monthly temperature from summer to winter in the lower levels throughout the islands is slight, ... while the drop in temperature from sea level to the highest elevations corresponds closely to what might be expected of temperature gradients in the tropics. . . . Generally speaking, the extreme highest temperatures have been recorded on the leeward, and, consequently, the drier, sunnier sides of the islands, where the slopes are presented more directly to the rays of the afternoon sun . . . . The higher levels of Mauna Kea (13,825 feet) and Mauna Loa (13,675 feet), Hawaii, and of Haleakala, (10,032 feet), Maui, are frequently white with snow during the colder months of the year, and it is not unusual to see snow banks on the crest of Mauna Kea or Mauna Loa, the highest peaks of the islands, even in midsummer. . . . . . It is doubtful if frost ever forms much if any below the 2,500-foot level over the entire group of islands, and rarely below the 4,000-foot level.

In summary, the strongly contrasted types of climate prevalent over different sections of the Hawaiian islands result from the presence of these land barriers, rising to various elevations above sea level, in the path of the prevalent northeast Trade Winds. Savanna climates prevail over the lower windward slopes of the higher islands. At higher elevations (between 2,500 and 6,000 to 7,000 feet) the rainfall is heavier and is more evenly distributed throughout the year. The climate is like that of the rain forest belt of the Tropics. On the highest mountains, above the belt of maximum precipitation the rainfall is less and the temperatures are lower and show greater daily and seasonal variability. The general features of tropical climates prevail, but temperature rather than rainfall is the controlling factor. These highland climates are isothermal and depend for their special characters upon elevation above sea level.

The leeward slopes of the higher domes receive lesser amounts of rainfall as sea level is approached; the climates vary from tropical highland isothermal at the highest elevations to savanna at intermediate levels and become semiarid near sea level. On Niihau (1,281 feet), dry climates everywhere prevail; the rainfall rarely exceeds 25 or 30 inches annually, and commonly is much less. The island is too low to cause the formation of permanent cloud caps over its summit.

Rainfall

The rainfall map of Kauai (fig. 2) shows that the precipitation rapidly increases from the shoreward margin of the northern and northeastern lowlands to the summit, where Waialeale (5,080 feet), near the center of the island is credited for a period extending intermittently from 1911 to 1917 with the extraordinary annual average rainfall of 476 inches (Table 3). Waialeale therefore rivals Cherrapunji, Assam (average annual rainfall 426 inches), as a claimant for the doubtful honor of being the rainiest place on earth. The precipitation over the windward lowlands is generally more than 50 inches annually, and decreases from the northern and eastern coasts

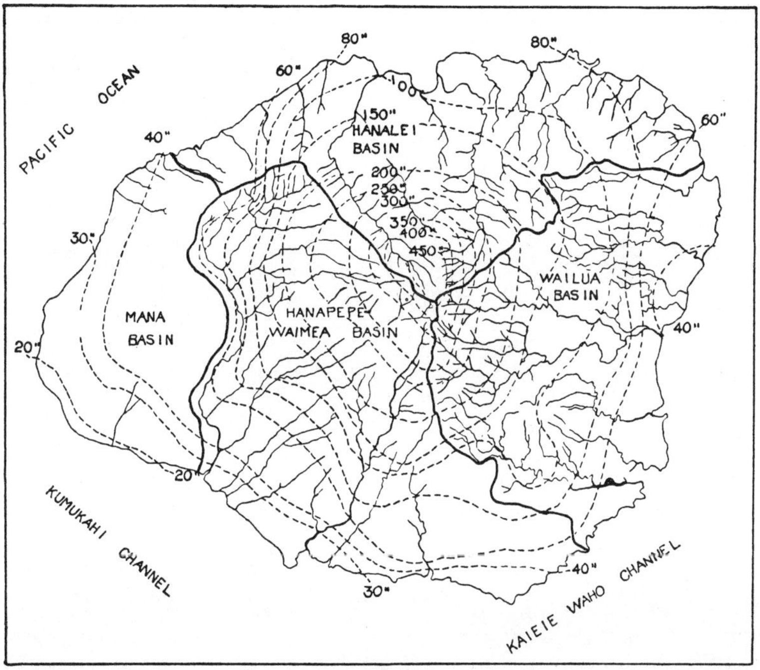

FIGURE 2.—Rainfall and drainage map of Kauai. Drainage from U. S. Geological Survey, Water-supply paper 318. Rainfall data from the U. S. Weather Bureau. Bounderies of drainage basins are indicated by heavy lines. Scale: 1 inch=8.85 miles.

toward the southern coast of the island. To the leeward of the summit bog, on which Waialeale is located, the rainfall rapidly lessens, and over the mountain slopes and coastal lowlands 10 to 15 miles from the center of the island, the climate is distinctly arid. Tables 1 and 2 (compiled from publications of the United States Weather Bureau) show the distribution of precipitation over Kauai. (See also fig. 2)

The rainfall records for Waialeale, the station at the summit of the island (elevation 5,080 feet), so far as they are available, are given in Table 3. These records do not include the greater part of the years 1914 and 1918, the wettest in the history of the local Weather Bureau Office. The mean rainfall for all periods covered by actual rainfall records was 1.30 inches a day, or, 476 inches a year. Unfortunately the station is very difficult of access and from time to time has been discontinued. When certain of the observations were made, the rain gage was found to have overflowed, so that the maximum precipitation at the station probably is not known.

The decrease in precipitation over the leeward slopes of Kauai is rapid but irregular, for strong air currents blow tongues of the cloud cap, which is so commonly present over the upper levels of the island, for some distance down the leeward canyons and over the intervening divides, and thereby

TABLE 1. RAINFALL OVER WINDWARD AND LEEWARD SLOPES OF KAUAI.

WINDWARD SLOPES					LEEWARD SLOPES				
Station	Distance from station of heaviest rainfall (miles)	Elevation (feet)	Average annual rainfall (inches)	Length of record (years)	Station	Distance from station of heaviest rainfall (miles)	Elevation (feet)	Average annual rainfall (inches)	Length of record (years)
1	10.0	10	98.50	18	8	0.0	5,075	476.00	7
2	10.5	105	89.47	8	9		3,600	94.85	6
3	10.0	125	124.91	11	10	11.1	2,550	55.73	6
4	11.5	342	70.48	33	11	5.5	3.500	46.58	5
5	7.5	635	145.83	8	12	10.0	900	67.56	12
6	8.0	700	184.09	12	13	13.0	850	15.81	7
7	4.0	1,900	198.59	8	14	13.0	150	29.36	18
8	0.0	5,075	476.00	7	15	16.5	35	22.21	17
					16	4.5	2,100	144.58	7
					17	1.5	2,080	237.05	9
					18	7.5	2,000	170.49	15
					19	4.0	1,310	111.65	14
					20	5.5	530	133.09	14

extend the belt of heavy rains beyond its normal limit. Thus may be explained the heavy precipitation recorded at leeward Stations 16-20, Table 1. Stations 16 and 19 are at or near the bottom of Olokele Canyon, Station 20 at the bottom of Hanapepe Canyon, and Stations 17 and 18 on neighboring divides. The rainy season which lasts from November through March on the leeward side and through April on the windward side, is well defined over the lowlands; at higher elevations, the precipitation is much more evenly distributed throughout the year. (See Table 2.) The mountain summits are cloud-capped, except for a few hours during the early morning or for a few days during occasional southerly or southwesterly winds. On the windward side, the base of the cloud cap generally hangs about 2,000 feet above sea level; on the leeward side, the base of the cloud is much more irregular and generally does not extend so far down the mountain sides. Within the area more or less continuously covered by the cloud cap, rain falls most of the time. That portion of this area comprising the summit plateau has low relief and relatively poor drainage, hence nearly all of it is swampy.

TABLE 2. RAINFALL BY MONTHS OVER COASTAL LOWLANDS AND MOUNTAIN SLOPES OF KAUAI. (IN INCHES.)

Station[a]	Elevation (feet)	Jan.	Feb.	March	April	May	June	July	Aug.	Sept.	Oct.	Nov.	Dec.
1	200	6.72	3.77	7.66	3.85	3.70	2.80	2.73	3.41	5.17	3.64	5.28	6.38
2	10	6.48	9.88	15.02	7.90	6.35	5.28	6.67	8.35	6.40	7.61	9.24	9.31
3	155	4.04	3.45	4.94	1.94	1.42	1.19	1.18	1.22	2.04	1.83	2.91	3.20
4	30	4.20	1.59	5.49	1.08	1.33	0.69	0.49	0.61	1.82	1.51	1.87	3.12
5	700	12.89	14.25	21.61	20.89	15.10	12.28	12.80	14.49	11.86	11.18	18.34	18.40
6	1900	18.02	10.28	19.98	17.80	18.04	14.05	15.61	13.56	16.97	14.29	21.05	18.94
7	2080	19.02	10.70	16.92	20.15	17.79	18.78	22.51	19.08	20.71	19.14	27.08	27.17
8	2100	12.72	7.83	12.44	12.06	10.82	9.53	10.49	10.30	12.02	12.24	17.50	16.53

[a] Stations 1, 2, windward coastal plains; 3, 4, leeward coastal plains; 5, 6, windward mountain slopes; 7, 8, leeward mountain slopes.

A striking feature of the Hawaiian rainfall is its large annual variability. (See Tables 3 and 4.) The Kona storms materially affect the annual precipitation, especially over the leeward slopes, as the precipitation during a single storm may exceed the annual average for the region. Table 4 gives the maximum and the minimum annual rainfall recorded at certain stations on Kauai, and a few maxima measured for 24-hour periods.

TABLE 3. ANNUAL RAINFALL AT WAIALEALE, KAUAI (IN INCHES).

1911		1915	
Aug.-Oct.	134.00	June-Sept.	197.00
Nov.-Dec.	86.00	Oct.-Dec.	217.00
1912 (Jan. 6, 1912-Jan. 11, 1913)		(Precipitation May 21, 1915, to May 21, 1916, 561.0 inches.)	
Jan.-Mar.	90.00		
Apr.-June	82.20		
July-Sept.	78.50		
Oct.-Nov.	58.00	1916 (Jan. 12, 1916-Jan. 4, 1917)	
Dec.	105.50	Jan.-May	147.00
		June-July	75.00
Total	414.20	Aug.-Oct.	109.00
		Nov.-Dec.	190.00
1913 (Jan. 11, 1913-Jan. 13, 1914)			
Jan.-Mar.	60.00	Total	521.00
Apr.-May	94.00		
June-July	77.00		
Aug.-Sept.	61.50	1917	
Oct.	23.50		
Nov.-Dec.	135.00	Jan.-Apr.	166.00
		May-Aug.	109.00
Total	451.00	1919 (June-Dec.)	204.00
		1920	549.00
1914		1921	367.00
Jan.-Mar.	30.00	1922	452.00
April-July	134.00	1923	360.00
(Gage found overflowing. Station discontinued till May 21, 1915.)		1924 (Jan.-July)	228.00
		1925 (Jan.-July)	362.00

TABLE 4. RECORDED RAINFALL AT CERTAIN STATIONS ON KAUAI.

Station		Elevation (feet)	Minimum annual rainfall (inches)	Date	Maximum annual rainfall (inches)	Date	Elevation (feet)	Rainfall 24-hour maximum (inches)
1	Windward lowland	10	64.68	1906	161.45	1902	140	9.52
2	Windward lowland	200	32.66	1912	71.50	1918	250	11.40
3	Leeward lowland	8	5.97	1912	38.71	1907	700	16.00
4	Leeward lowland	150	14.35	1912	43.17	1907	1055	10.52
5	Windward mountains	700	126.54	1908	249.83	1916	3500	15.10
6	Windward mountains	1900	155.82	1912	237.35	1918	3723	24.40
7	Leeward mountains	2080	189.23	1923	303.20	1906	....	....
8	Summit of island	5080	367.00	1921	549.00	1920	....	....

Temperature

The only temperature records available for Kauai are from six stations on the lowlands. These stations, of course, show small differences between the mean monthly maxima and the mean monthly minima. The mean average temperatures are slightly higher on the leeward than on the windward side of the island. (See Table 5.)

TABLE 5. TEMPERATURES AT STATIONS ON KAUAI.

STATION		MEAN ANNUAL TEMPERATURE RANGES	MEAN ANNUAL TEMPERATURE	COLDEST MONTH	WARMEST MONTH
Kilauea	Windward	68.5 - 75.3	72.4	January	July
Kealia	Windward	62.2 - 77.2	73.5	January	July-August
Lihue	Windward	68.0 - 76.5	72.5	January	July
Koloa	Leeward	69.2 - 76.7	73.1	January	September
Makaweli	Leeward	71.0 - 78.3	74.9	January	July
Mana Pump	Leeward	69.3 - 77.8	73.7	January	July

The temperatures over the highlands decrease, month for month, as elevation increases. On the mountain slopes, above 3,000 feet, frost is not uncommon during the rainy season; snow has not been observed. Unfortunately, no meteorological records are available regarding the temperatures in the highlands of Kauai, though they are probably like those at upland stations on other islands. On Maui, the mean temperature ranges at Keania Valley (altitude 1,600 feet) is 65.0° to 70.7°; at Kula Sanitarium (altitude 2,700 feet), 60.4° to 66.8°. On the island of Hawaii, the mean temperature at Volcano Observatory (altitude 3,984 feet) is 57.8° to 62.7°; at Humuula (altitude 6,685 feet) 48.5° to 55.5°.

Relation of Relief to Clouds and Rainfall

The relief of eastern and northeastern Kauai exerts an important influence upon the present distribution of rainfall over the windward slopes. The northern and eastern mountains rise abruptly from the lowlands to elevations of 2,000 to 5,100 feet. The moisture-laden trades, sweeping across the low country, are confronted by this barrier, which forces them to rise suddenly to levels of lower temperatures. Continuous condensation of the moisture follows, and a cloud cap is formed over the highlands above an elevation of 2,000 or 2,500 feet during most of the time that the Trade Winds blow. The position of the cloud cap is apparently controlled by the average temperature and the strength of the wind. On warm days, when the temperatures over the lowlands exceed 85°, the clouds rise from 1,000 to 2,000 feet above their normal level, and cover only the highest summits of the dome. At times when the wind velocity falls below the normal 10 to 12 miles an hour, the cloud cap rises or disappears entirely for a few hours. When the occasional southerly or southwesterly winds prevail, the cloud cap is generally dissipated for periods of a few days, the longest intervals during which the highland summits are visible.

The condensation of the water vapor which takes place when the Trade Winds are forced to cooler levels not only results in cloud formation, but also produces more or less continuous heavy precipitation over the higher windward slopes and over the summit plateau. Because of this suddenness of the upward deflection of the air currents, the increase in precipitation along the mountain slopes is rapid, as is shown in Table 6.

TABLE 6. DISTRIBUTION OF PRECIPITATION OVER NORTHERN AND EASTERN KAUAI.

STATION	ELEVATION (FEET)	AVERAGE ANNUAL PRECIPITATION (INCHES)	LENGTH OF RECORD (YEARS)
Northern Kauai:			
Outer portion of coastal lowland			
1	10	98.50	18
2	105	89.47	8
3	342	70.48	33
Inner margin of lowland			
4	635	145.83	8
Canyons and mountain slopes			
5	125	124.91	11
6	700	184.00	12
7	1900	198.59	8
8	5075	473.00	7
Eastern Kauai:			
Outer portion of lowland			
9	14	40.81	19
10	200	55.11	14
11	350	65.53	9
12	400	60.15	16
Inner margin of lowland and lower mountain slopes			
13	500	119.31	3
14	650	125.85	4
15	725	91.84	14
16	835	125.18	11
17	911	108.85	3
Higher mountain slopes			
18	1900	198.59	8
19	5080	473.00	7

Hamrick (29) has discussed the causes of the formation of these cloud caps on the Koolau Mountains, Oahu, which closely resemble the steep slopes of northern and eastern Kauai. On theoretical grounds he placed the base of the cloud cap at 2,040 feet, a figure which closely approximates the actual conditions.

Because of the wide plains on windward Kauai and eastern Oahu the rainfall is less at a greater distance from the shore than on the other domes of similar or greater elevation, and the rate of increase of precipitation is more rapid over the very steep windward slopes of these mountains than over those having the more normal, gentler slopes.

GEOMORPHOLOGY

FORM AND SIZE OF KAUAI AND NIIHAU

The islands of Kauai and Niihau are the highest summits of one of the principal volcanic mountains of the Hawaiian range, and represent accumulations about the two major eruptive centers which have appeared above sea level. Soundings may reveal submerged centers, especially off the southern end of Niihau, where comparatively few depths have been measured. The greater portion of the mountain, of course, is submerged; the depths of the Pacific surrounding its base range from 2,000 to 2,500 fathoms, whereas the highest elevation on Kauai is slightly over 5,000 feet and on Niihau about 1,300 feet above sea level. The area of the base of the mountain cannot be accurately determined; but the long diameter, if the cinder cone, Kaulu, 19 miles southwest of Niihau be a part of the structure, must exceed 100 miles and the short diameter 40 to 50 miles. The major and minor diameters of Kauai are 32 and 22 miles respectively and of Niihau 18 and 5 miles. Niihau was originally larger and higher than at present; marine abrasion and downfaulting have caused the disappearance of more than half of the island. A discussion of the surface relief applies therefore to only a small part of the volcanic edifices. Little is known of the structure of the submerged part of the mountain. (See p. 50.)

FEATURES OF HAWAIIAN LAVA DOMES

In this report, I shall use the term "lava dome" as defined by Daly (11, pp. 135, 136):

> The greater masses of lava, which in the form of many individual flows, have issued from a central vent [or, as Daly (14) has demonstrated since, from a series of more or less centrally located principal fissures] in the proper abundance and proper directions to build a dome-shaped pile of lava . . . . Such bodies are of exogenous growth and are thus contrasted genetically as well as in size with the plug-domes.

Von Wolf (64, p. 448) calls such mountains "Lavavulkane" or preferably "Schildvulkane," and recognizes two classes: (1) the Dyngju type of Reck and von Knebel which is found in exceptional abundance in Iceland; (2) the Hawaiian type. As size is the sole basis of this classification, the distinction seems unnecessary; the smaller Hawaiian domes, west Molokai and Kahoolawe, do not greatly exceed the dimensions of some domes of the Dyngju type. The coalescence of two or more of these lava domes forms a compound lava volcano; the various mountains may be joined together above sea level as are Hawaii (built of at least five domes), Oahu and Maui (each built of two domes); or they may be connected below sea level as are Maui, Molokai, Lanai, and Kahoolawe, or Kauai and Niihau. It seems highly probable that the Oahu doublet has been joined with the great compound Maui volcano below sea level to form a single mountain of huge dimensions and mass (36, pp. 149-150.)

At the close of their last principal eruptive periods, the Kauai and Niihau domes probably resembled in form the young volcanoes of Hawaii: Mauna Loa, Kilauea, Mauna Kea, and Hualalai.

As typified by Mauna Loa and Hualalai, the completed or nearly completed lava dome is a land form of simple outline, composed of more or less homogeneous layered rocks which dip seaward at low angles from the major eruptive area or center or from other openings along the flanks of the volcano. The changes in relief which follow the cessation of volcanic activity are produced chiefly by fluvial and marine erosion, by the gravitative transfer of rock or of the soil cover, and by faulting. Changes of level also may be of importance and the sporadic renewal of volcanic activity at local centers frequently adds new lavas and deposits of ejectamenta to the erosional surface. In colder latitudes, glaciation may modify the pre-existing relief.

Comparison with Mauna Loa and Hualalai indicates that at the close of the last principal eruptive periods Kauai and Niihau were somewhat symmetrical, gently sloping mountains, on whose surfaces were imposed numerous parasitic cinder cones. Permanent valleys were few and unimportant, and sea cliffs, if existing, were low. The topography of the central eruptive area cannot be determined on either island. The principal center on Kauai is either concealed beneath the thick jungle cover over the highlands or has disappeared wholly or in part as the result of the downfaulting and deep erosion of a large section of the eastern part of the dome. Engulfment and marine erosion have caused the disappearance of the major eruptive center on Niihau.

The height of Kauai at the close of the last major eruptive period probably was about the same as at present, slightly more than 5,000 feet, the maximum elevation on the largest remnant of the constructional surface. If any reduction in height has taken place through erosion, crushing of the vesicular and cavernous lavas, or subsidence, this loss probably has been compensated by an emergence of the island which has exceeded 500 feet. The periphery on the other hand has been reduced by wave attack. Judging from the extent of the sea cliffing, the original maximum diameter of the island must have been approximately 40 miles, whereas the present greatest diameter is 32 miles. Niihau has lost notably in size and elevation; at the close of the principal eruptive period, the dimensions of the volcano probably were about 15 by 10 miles and its elevation at least 2,000 or 3,000 feet.

PHYSIOGRAPHIC SUBDIVISIONS OF KAUAI

The initial, simple, dome-like contour of Kauai has been greatly modified by long continued fluvial and marine erosion, by extensive downfaulting, by changes of level of considerable magnitude, and by explosive volcanic eruptions which took place from subordinate vents after most of the present topographic features had been developed. Four major divisions of the present highly diversified relief may be recognized (fig. 3): (1) The summit plateau; (2) the dissected highland; (3) the elevated wave-cut platform of eastern Kauai; (4) the constructional plain lying west of the Waimea River.

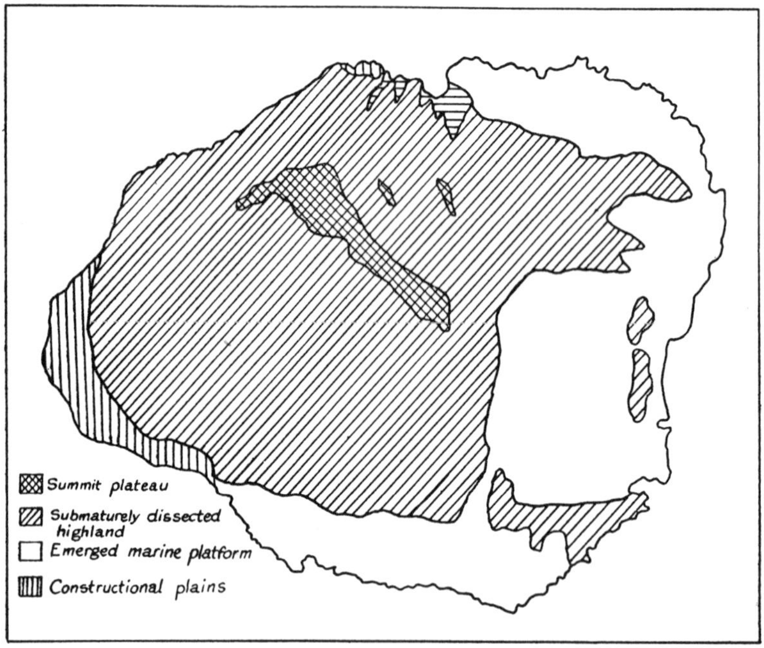

FIGURE 3.—Physiographic subdivisions of Kauai. Scale: 1 inch=8.8 miles.

SUMMIT PLATEAU

The summit plateau (Alakai Swamp) is the principal remnant of the constructional surface of the Kauai dome. This sector, which has an area of nearly 30 square miles (about one-twentieth of the island), extends for 10 miles northwestward from the center of the island; its northwestern boundary is 2.5 to 3 miles from the coast. The deep gorge of the Koaie River flowing southwestward from the plateau nearly bisects it. Northwest of this gorge, the average width of the plateau is about 2.5 miles; to the southeast, the average width is about 1.5 miles. The elevation of the sector decreases from 5,000 feet along its eastern margin to 4,000 feet along its northwestern margin. On the northwest, northeast, east, and south, the limits of the plateau are sharply defined by great palis, the cliffed walls and heads of the major canyons. On the southwest side, the upstream growth of the canyons has been slower and more irregular, owing to the lesser precipitation over the leeward slopes of the dome; hence the southwest margin is ill defined. There is, however, a rather sharp change in slope between this sector and the deeply eroded portion of the highland to the west. As is shown by the direction of flow of most of the streams, the general slope of the plateau is southwest. The eastern extremity slopes north and east, and from this section flow the headwaters of the principal rivers of northern and eastern Kauai.

The surface of the plateau is traversed by ill-defined drainage lines, which have eroded broad, shallow valleys. Northwest of the deep Koaie gorge, these valleys are somewhat deeper than to the southeast, apparently because of the somewhat higher surface gradient of the former area. Broad, tabular interfluves separate the valleys. Though the topography of the region is youthful, the valley forms are those developed in an upland region of low slope where downward erosion by the small, headward tributaries of the principal streams of the island is relatively slow. Beyond the limits of the plateau, valley cutting has proceeded much more rapidly, hence there is a striking contrast between the gentle rolling topography of the plateau and the highly differentiated relief of the rest of the highland (Pl. I, *B*).

On the summit plateau, rain falls almost constantly. (See pp. 15-16.) Because of the low relief, drainage is slow, and large areas of the surface are swampy. Low ridges and eminences standing above the general surface are better drained, but, even on these, water stands more or less permanently in every depression. Except on these higher areas, the plateau is practically treeless; the areas separating the pools of water are covered with grass, sphagnum mosses, lichens, sedges, a few shrubs, and other types of small plants. Jungle forest, similar to that of the rainier section of the dissected highland, covers the higher elevations of the plateau. Owing to the extent of the swamp and to the almost ever-present cloud blanket, travel over the plateau is difficult. Field observations regarding the morphologic and structural relations are difficult to procure; exposures of undecomposed rocks are absent on the summit plateau, and heavy vegetation covers the surrounding, more dissected areas where sections of the lavas might be found.

The summit plateau is a remnant of the constructional surface of the Kauai dome. Though the structural relationships cannot be definitely ascertained owing to difficulties of exploration, it is evident that the surface of the area in a general way lies parallel to the bedding of the lavas exposed in the gorges and canyons of the adjacent, more dissected portions of the highland. Also it is impossible to suggest any other way in which so extensive an area of low relief could be developed near the summit of a deeply eroded volcanic mountain. From the geomorphic evidence, therefore, it is apparent that the headward advance of the streams, especially on the leeward side of Kauai, is not yet completed, and that a small area of the original surface, only slightly modified by erosion, has been preserved. Eventually, the upstream growth of the cliffed canyons will reduce the central portion of the island to a series of sharp peaks, such as are present in many of the older, still more deeply eroded lava domes in various parts of the world.

Other smaller remnants of the constructional surface are preserved as seaward-sloping areas of low relief forming parts of the divides between certain of the canyons of northern and northeastern Kauai. While I was unable to visit any of these smaller remnants, it was apparent from the descriptions given by others and from distant views that their topography is similar to that of the summit plateau.

Though water stands in practically every depression in the plateau surfaces, little vegetation accumulates; the average high temperatures cause rather rapid decay of most of the dead plant tissues. Organic acids are thus added abundantly to the subsurface waters which percolate downward through the porous and cavernous lavas.

Dissected Highland

At the close of the principal volcanic period few valleys of any consequence existed on Kauai. The frequency of eruptions was sufficient to prevent active stream erosion. With the cessation of volcanic activity, waters, penetrating through the lavas, gradually decomposed them, and the surface run-off, concentrated in many depressions, began to incise narrow gorges into the mountain slopes. Assuming a distribution of rainfall similar to that of the present, the heavier precipitation over the windward section of the island caused the formation of a greater number of streams than on the leeward side. Near the summit of the dome, the streamlets were unable to materially corrade the lavas. The elevations at which valley cutting was initiated were in general higher on the windward than on the leeward flanks of the island, because of the more active weathering of the lavas, and the greater numbers and volumes of the streams. The valleys at first were shallow, steep-walled trenches, between which were practically undissected conic sections of the constructional surface. It is probable that most of these valleys did not enter the ocean at grade because in the early erosional stages, wave cutting apparently is more effective than stream erosion. For example, on windward Mauna Kea, one of the most recently extinct Hawaiian volcanoes, the streams tumble a hundred or so feet down sea cliffs into the ocean. As erosion progressed, the windward streams rapidly deepened their channels to grade. Along the leeward coast, downcutting has been so much slower that the mouths of many of the gorges still hang some distance above the base of the great cliffs.

The early erosional topography of Kauai may be compared with that of eastern Maui (Haleakala) or of Mauna Kea and Hualalai, all recently extinct volcanoes.

In the zone of heaviest rainfall on the windward slopes of Kauai, the multiplication of tributaries has gradually produced more highly complicated drainage patterns and consequently finer textured topography. The dissection of the seaward portions of the interfluves has taken place more slowly. On the windward side, the crests are sharp throughout their entire length; on the leeward side, they are ridge form in the interior of the island, and gradually become tabular near the outer margin of the highland (fig. 4). The topography at equivalent elevations on the leeward side is distinctly coarser textured. Study of East Maui and Mauna Kea shows that the asymmetrical

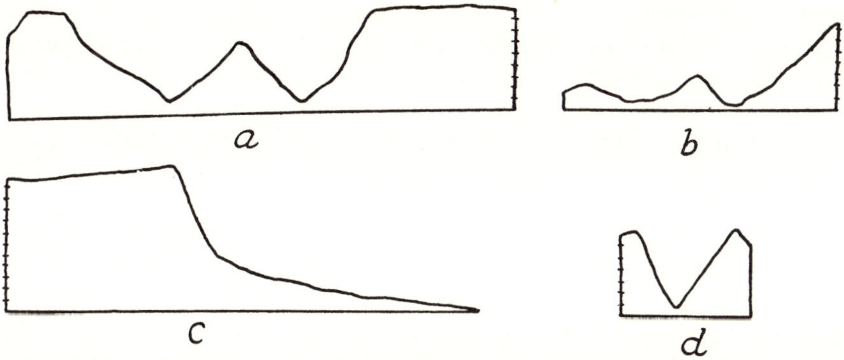

FIGURE 4.—Cross-sections (*a, b, d,*) of valleys and interfluves and profile of valley (*c*) of northern Kauai. Vertical scale: 1 space=250 feet; approximate horizontal scale: 1 inch=2 miles.

evolution of the erosional relief on the windward and leeward sides of mountains located in the belts of savanna climates begins in the early posteruptive stages; on the older domes it remains a striking feature in spite of the extensive denudation of all sections. On Kauai, this asymmetry has been accentuated because of two special conditions: (1) an abnormal development of the drainage on the western side of the island due to the deflection of a part of the streams by a fault zone along which the Waimea River runs (p. 30); (2) the downfaulting of a large section of the eastern part of the dome (p. 30).

Normally, a consequent "wheel-spoke" drainage and valley or canyon system is developed on conical mountains such as explosive volcanoes or lava domes. The major drainage lines on Kauai flow from the summit plateau and the immediately adjacent ridges, and follow roughly radial paths to the shore line. On the western side of the island, the north-south Waimea fault zone

(p. 30) existed, and apparently the block to the west stood higher than that to the east of the zone of dislocation. The streams from the heavily watered summit plateau were deflected by the scarp, and became tributary to a major channel (the Waimea River), flowing southward to the ocean (fig. 8, *a*). To the west of the river, the drainage again is radial. Other abnormalities, probably due to faulting, are to be found in some of the drainage lines of eastern Kauai.

The major streams and their principal tributaries have incised deep, narrow canyons into the heart of the dome, and, in the long course of the post-volcanic period, have aided in the very extensive destruction of the constructional surface. The stage of dissection naturally is most advanced on the windward (northern and northeastern) sides of the island; the minimum effects of stream erosion are to be observed west of Waimea Canyon, owing to special conditions.

Above the heads of the great canyons, the tributaries, flowing over the gently sloping constructional surface, have not greatly eroded it; and a small, only slightly dissected remnant of this surface has been preserved as the summit plateau of the island. Over most of the plateau, the valleys are inconspicuous. On the northwestern, eastern, and southern sides, the streamlets tumble down the steep-walled or cliffed heads and sides of the canyons into the trunk streams below. On the southwestern side, toward which most of the plateau slopes, the headward growth of the major canyons has been slower. The plateau drainage has united into a number of streams powerful enough to cut steep-walled gorges several hundred feet deep into the margin of the area but these are much shallower than the principal canyons which they join. The abundance of these hanging valleys is one of the most striking features of the relief of Kauai and the other dissected lava domes in Hawaii (Pls. I, *B;* II, *B*). Thus, on the leeward side, there is a transition zone between the relatively undissected summit plateau and the belt of the great canyons to the southwest (Pl. I, *B*) ; this difference in the evolution of the erosional relief is of course a reflection of the asymmetric distribution of rainfall over the windward and leeward sections of the island.

The continued headward growth of the main canyons will eventually result in the destruction of the summit plateau; the central portion of the island will then consist of a group of jagged peaks instead of the flattish area now present.

The canyons are narrow, roughly radial trenches, 1,000 to 3,000 feet in depth, a maximum of 10 or 12 miles in length, and 1 or 2 miles in average width; their heads are huge, steep-walled amphitheaters which bear some resemblance to glacial cirques in their rounded ground plan, the general sheerness of their walls, and the steepening of the upper portions of their walls. A few valleys even have flattish floors (fig. 4). Freeman (25) has suggested the term "pseudo-cirque" for such amphitheaters, but I have objected to this on the ground that it is undesirable to apply terms definitely associated with certain geological process to forms evolved by a totally different process (35). The name "amphitheater head" or "amphitheater valley head" is satisfactory.

In the rainy sections of the highland, tributary streamlets, plunging from the summit plateau or from the sharp-crested interfluves, have deeply channelled the canyon walls, and have produced a highly complex "organ-pipe topography." Specially resistant lavas within the igneous complex have not been cut back by the streams as rapidly as weaker flows, hence the waters, in their descent into the major drainage lines, commonly fall on to a number of narrow rock benches. Into these benches, the streams have cut shallow basins and have undercut to some extent the rocks immediately above the benches. The furrows through which the falls pass are deep; between them are huge, nearly vertical columns or pipes, many of which in turn have been fluted or divided into smaller pipes by minor streamlets. Some of the falls are permanent, others appear only after heavy rains, but water flows through the channels sufficiently often to prevent the growth of much vegetation over the lavas. Hence the fall lines appear as barren streaks of black rock through the dense jungle tapestry which covers practically all of the canyon walls in the rainy section. During and after heavy rains, in nearly every furrow is a silvery ribbon of water tumbling for hundreds of feet through the green of the forest cover, its fall broken at various levels by projecting lava benches or masked for a few feet by a specially thick tangle of vegetation. The charm of the tropical landscape is thereby increased a hundred fold. (See Pl. III, *C*.)

In the drier, lower reaches of the canyons, minor streams are fewer, and the "organ-pipe" topography gradually disappears. High, nearly vertical cliffs form long stretches of the canyon walls (Pl. III, *A*). Cliff-bench profiles, caused by the difference in resistance of the lavas to erosion, are common. Vesicular ropy flows and highly brecciated clinkery flow tops form weak layers along which undermining proceeds with maximum rapidity. Above these layers collapse of insufficiently supported rock masses occasionally takes place. In Waimea Canyon in particular, there is so marked a development of cliff-bench topography that this gorge is frequently called a miniature Grand Canyon. Viewed from its western rim, the canyon does bear a rather striking resemblance to the master gorge, though of course it lacks the incomparable coloring (Pl. I, *B*).

The depth and narrowness of the canyons and the steepness of their walls result from the speed at which down cutting has taken place and the average high resistance of the lavas to erosion. At the surface, the lavas are deeply weathered, and the streams rapidly cut through this oxidized zone to fresher rocks below. Frost does not form sufficiently often in the highlands of Kauai to be an important agent of rock disintegration; at lower elevations freezing temperatures never occur. Changes of temperature also are slight even at the highest elevations, and have virtually no effect upon the rocks. Because of the abundant precipitation, the average high temperatures, and the extremely porous structure of the lavas, chemical decay proceeds rapidly; atmospheric and ground water solutions readily penetrate and attack the rocks, and, on the windward side of the island in particular, have altered them to considerable depths. The highly vesicular and brecciated zones within the flows provide avenues of specially easy access for the subsurface circulation; hence, where rocks possess these structures, weathering has proceeded rapidly. As the general peripheral dip of the flows together with the great numbers of openings in them allows for the active migration of underground waters, the frequent acquisitions of oxidizing solutions are readily taken care of. Rapid decay of abundant vegetation in the warm, humid climate adds great quantities of organic acids to the solutions, and increases their efficiency as agents of rock weathering especially in the surface zones. On the summit plateaus and on the windward side of Kauai, decomposition of the lavas had proceeded far below the surface; on the leeward side, the depth of the oxidized zone is much less. Mechanical disintegration is at a minimum, except in the channels of the streams, along the shore lines, and in places where landsliding of large rock masses takes place.

The major streams are large and have moderately steep gradients; hence they are rapidly deepening their canyons and eroding them headwards. Sapping by weathering and running water takes place actively along weak zones in the lavas. Landsliding is common and is the principal factor in the widening and headward growth of the valleys, the steepening of their walls, and the supplying of rock fragments to the streams. In the areas of heavy rainfall, the abundant and speedy run-off normally should cause active solifluction and should keep the slopes relatively free from loose material, but even the steepest slopes are covered with dense vegetation which retards the movement of soil and also is effective in preventing the slump of larger, more or less unstable masses of weathered rock. Solifluction is greatly accelerated during the rainy season, and the swollen streams carry tremendous quantities of detritus washed into them from the steep slopes. Frequent landsliding of course takes place even where the vegetation is heaviest; great scars along the canyon walls record such collapse. Where the vegetation is luxuriant, these scars are quickly obscured. In the drier sections, the rate of weathering is slower and sapping along weak zones takes place less rapidly. The plant cover is broken, hence there is little to check the slump of loose débris, but the amount of material available for transfer is infinitely less than in the rainy sections.

SUBDIVISIONS OF THE DISSECTED HIGHLAND

Because of certain marked topographic differences, the highly dissected portion of the island surrounding the summit plateau may be divided into five segments or parts of segments (fig. 3 and p. 45) as follows:

1. The segment between the western rim of Wainiha Canyon and the Anahola Mountains—the eastern part of the Hanalei drainage basin.
2. The area between the Anahola Mountains and Haupu Ridge—the mountainous section of the Wailua drainage basin.
3. The segment between Haupu Ridge and the western rim of Waimea Canyon—the Hanapepe-Waimea drainage basin.
4. The area west of Waimea Canyon and south of Nualolo Canyon, northern Kauai—a part of the Mana drainage basin.
5. The area between the west rim of Wainiha Canyon and Nualolo Canyon, northern Kauai—the western portion of the Hanalei drainage basin.

1. The most heavily watered section of Kauai is the rugged, submaturely dissected highland belt on the northern side between the west rim of Wainiha Canyon and the eastern end of the Anahola Mountains. West of the Kilauea River, the elevations of many of the mountain summits range from 3,000 to 5,000 feet; at the east end of the Anahola Range, they decrease to about 2,000 feet. The heaviest precipitation falls over the western part of the segment; in that part are the largest streams on the island, streams which have the greatest number of tributaries (fig. 5) and fluctuate least in their volumes. Though the topography of the whole area is submature, the difference in rainfall and in the size and number of the streams is reflected in the much greater dissection of the western half where the most stupendous canyons on the island are located. The larger of these great gorges have been eroded headward almost to the center of the dome. The interfluves are steep and narrow, and are deeply furrowed by scores of streamlets plunging down their sides; the crests are sharply serrated throughout their length (Pl. IV, *A*), except for occasional small, flattish areas which are remnants of the original constructional surface of the dome (fig. 4). In the western part of the segment, the canyon walls are covered by an almost impenetrable jungle tapestry; to the east, where less rain falls, the vegetation is not so dense.

At the base of the mountains east of Hanalei River is a wave-cut platform in places 3 and 4 miles in width, standing about 500 feet above sea level along its inner margin (Pl. IV, *B*). The streams which formerly flowed into the ocean at the mouths of the canyon have been extended across this plain and have cut deep, narrow trenches into it.

2. In the segment of eastern Kauai between the Anahola Mountains and Haupu Ridge, the development of a great volcanic sink has materially affected the topographic evolution, and has produced striking asymmetry in the east-west cross section of the island (figs. 6, 7). The magnitude of the

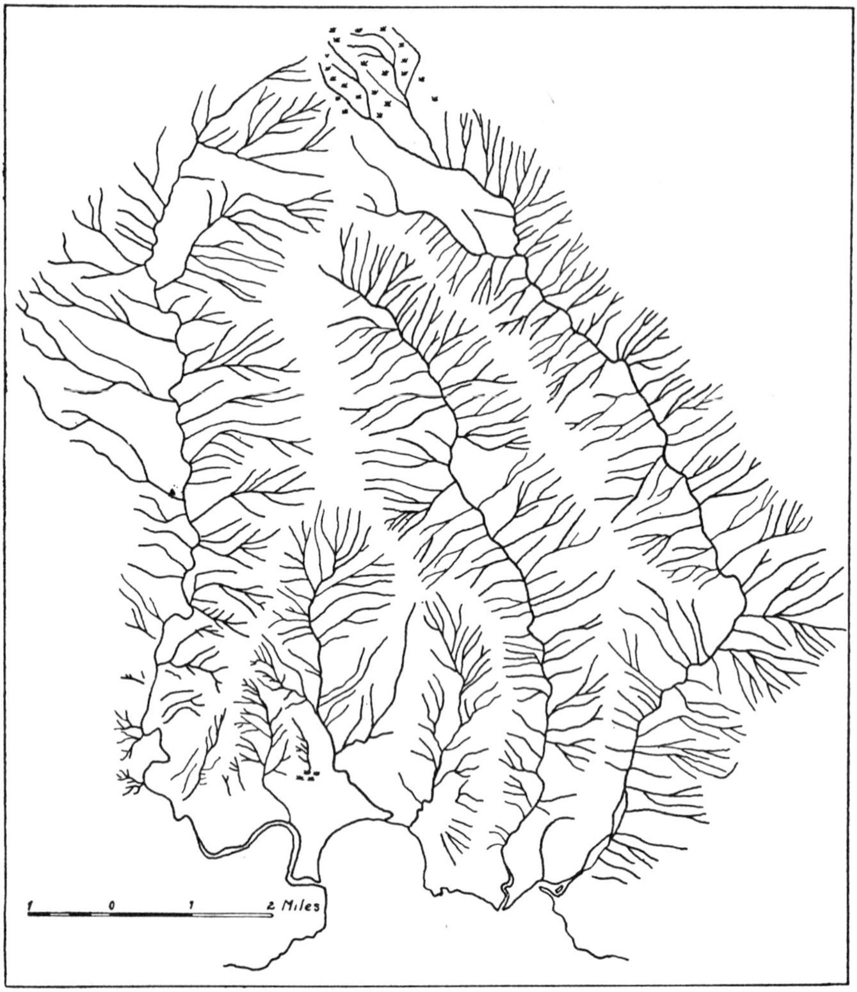

FIGURE 5.—Drainage map of a portion of northern Kauai.

original engulfed mass cannot be determined because subsequent erosion has greatly modified the whole of this area. Streams have gnawed huge canyons into parts of the walls, the floor of the depression has been bevelled off by waves, and the wave-cut platform has been elevated so that its inner margin now stands about 500 feet above sea level. At present, the floor of the Kauai sink measures 6½ miles in an east-west direction and 9 miles in a north-south direction; the dimensions from rim to rim are 10 miles east-west and 12 miles north-south. The highest point on the western wall is about 4,000 feet above the base, on the northern wall about 1,600 feet, on the eastern wall about 1,000 feet, and on the south wall about 2,000 feet (fig. 7).

Due to this engulfment, the great canyons bordering the summit plateau are short as compared with those in the segment first described, but have been eroded to similar depths. Before the emergence of the marine platform, the streams ran directly into the ocean, but now they flow for most of their lengths in valleys eroded into the wave-cut structure. Since the rain-

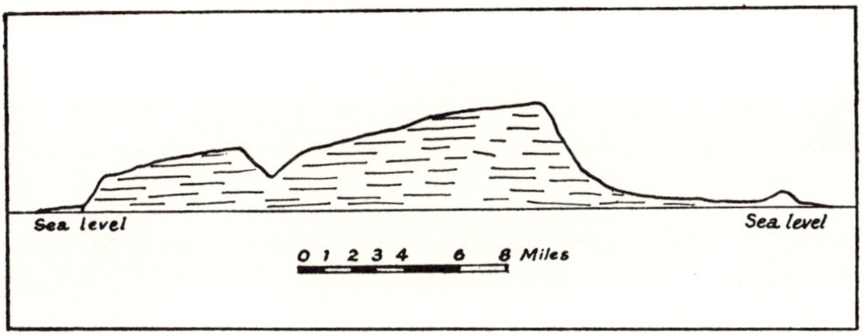

FIGURE 6.—East-west cross-section through the island of Kauai. Vertical scale: 1/10 in. = 1,000 ft.

fall over the higher slopes of this segment is only slightly less than over the adjacent segment to the north, the development of the drainage systems and character of the erosional relief in these two parts of the dissected highland is similar (Pl. IV *C*).

The parts of the highland now separated from the main area, Haupu Ridge on the south and the two lower ridges paralleling the eastern shore (Nonou and Kalepa), receive much less rain, and consequently have suffered relatively little erosion by streams (Pl. V, *B*).

3. The area lying between Haupu Ridge and Waimea Canyon is on the southern, leeward side of Kauai; most of it therefore is less heavily watered than either of the segments previously described. The precipitation is very heavy over the higher elevations, but decreases to 30 or 40 inches along the sea coast; also, more rain falls over the eastern than over the western part of the segment. The reflection of the lighter precipitation is shown in the less advanced stage of the erosional relief, as compared with that of highland of northern and eastern Kauai. The greater part of the summit plateau slopes to the southwest, and most of the run-off from its surface flows across this segment to the ocean (fig. 2). The major streams therefore have eroded huge canyons, but, since the volume and erosive power of the streams are less

than in the windward areas, the canyons are not so large nor has their headward growth proceeded so far toward the center of the island (Pls. I, *B*; III, *A*). In the eastern part of the area, the relief is more highly differential than in the western part because of the heavier rainfall and the more numerous and larger streams. The streams, having their source below the

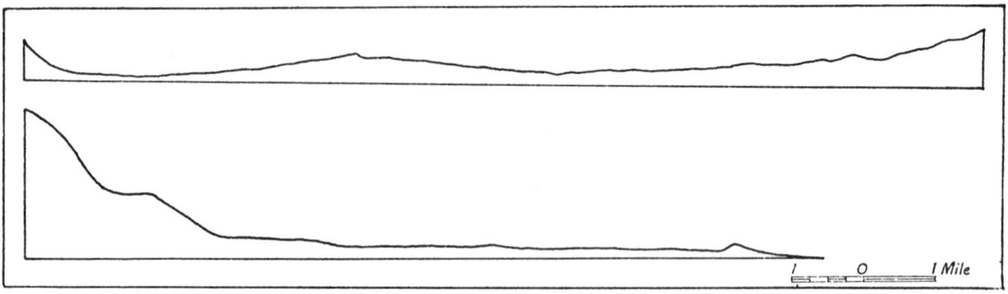

FIGURE 7.—East-west and north-south cross-sections through the volcanic sink, eastern Kauai.

margins of the summit plateau are small, and many of them are intermittent; their gorges also are small as compared with those of the main rivers.

In the higher portions of the leeward mountains, the drainage pattern is complex and the divides between the streams are steeply sloping, narrow crested ridges (Pl. III, *B*). Toward the sea coast, where the tributaries are fewer and erosion slower, the interfluves are broad and tabular in form.

The most striking feature of the segment is the Waimea drainage. The streams flowing from the summit plateau toward the western and southwestern sides of the island have been deflected by a north-south fault which has caused them to become tributary to a master channel following the general course of the dislocation. The drainage pattern is distinctly asymmetrical, since practically all of the tributaries enter from the eastern side (fig. 8, a). The trunk stream and each of its principal branches have cut great, cliff-walled canyons, into which open many hanging valleys of tributary streams (Pl. I, *B*). Since these streams have smaller volumes than those flowing from the summit plateau in the eastern part of the segment, their canyons are also smaller in size, and have not been eroded headward so far toward the center of the island. A number of small streams, heading below the summit plateau, also are tributary to the Waimea River. These have eroded short but deep cliff-walled gorges. Along the outer margin of the area, the uplift of the wave-cut platform has caused the extension of drainage which formerly flowed directly into the ocean from the mouths of the canyons.

4. As none of the drainage from the summit plateau reaches the western and southwestern coasts, the streams between the Waimea River and Nualolo Canyon derive most of their water supply from the scanty rains which fall

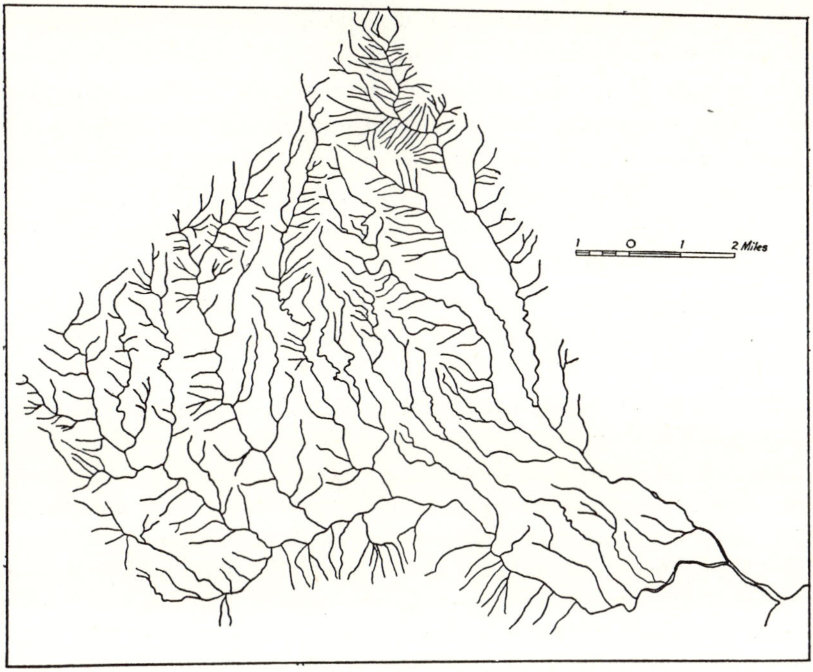

a

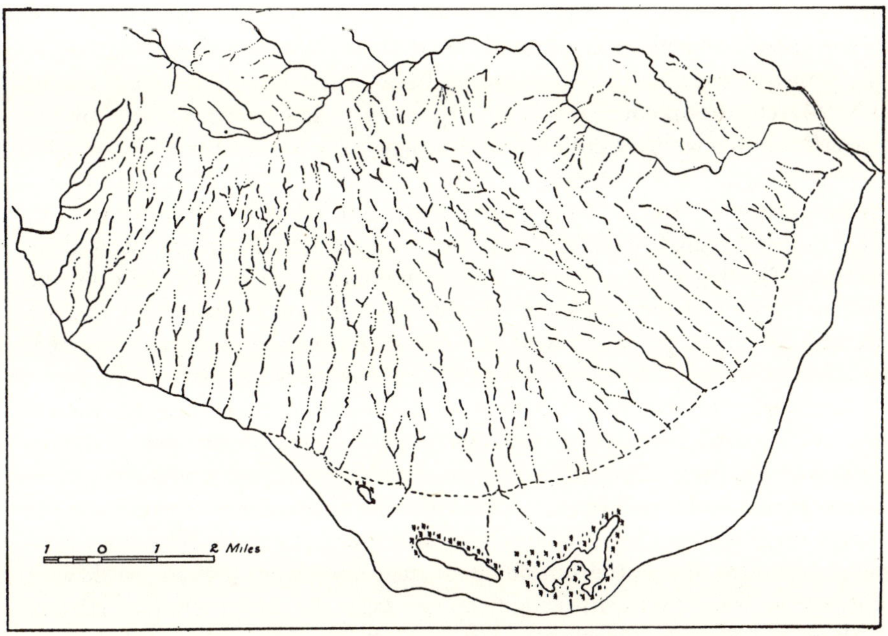

b

FIGURE 8.—*a,* The Waimea drainage system, southern Kauai; *b,* drainage systems in the Mana Basin, western Kauai. North is toward the left. (See plate xiii.)

over the area (fig. 8, b). The streams are small, and most of them intermittent; they flow mainly during the rainy season, when raging torrents pour down the gorges for a few days or a few weeks at the most. Erosion therefore has been slower than elsewhere on the island, and the topography of the area is yet youthful. Toward the sea coast, the streams have cut narrow, deep, frequently cliff-walled gorges, which generally are not more than 2 miles in length. Farther inland, the valleys of the less powerful tributaries hang many hundreds of feet above the floors of these canyons. The interfluves of the inland portion of the area are ridges of moderately steep slopes; toward the coast, slightly rounded or tabular areas, generally not more than a mile in width, separate the canyons. While the surfaces of these interfluves are not remnants of the constructional surface, most of them have been lowered only slightly below this initial level.

5. Northwest of the summit plateau between Wainiha and Nualolo streams, the canyon topography resembles that of the area between the Anahola Mountains and Haupu Ridge, eastern Kauai (segment No. 2) except that the stage of erosion is not so advanced. The canyons do not exceed 3 miles in length, but are very deep, narrow alcoves, separated by narrow, extremely precipitous, sharply serrated ridges (Pl. III, *C*). The largest canyons have been cut by streams receiving their water supply from precipitation over the summit plateau. The rainfall over the region is heavier on the eastern side than on the western, hence the drainage patterns of the former area are more complicated and the canyons in general are larger. The principal streams of the eastern part have cut the mouths of the gorges to sea level; the minor streams on the eastern and practically all on the western side plunge over cliffs of various heights into the ocean (Pl. II, *A*).

The reason for the shortness of the canyons in this section is not altogether apparent. Dana (19, pp. 278, 279) and Powers (52, p. 514) consider that the straight coast line of northwestern Kauai has been caused by the down-faulting of a peripheral slice of the dome. If this be the case, the canyons have undoubtedly been truncated by this dislocation and thereby reduced in length. Positive evidence of such faulting was not obtained, though a pronounced break in the submarine contours suggests it.

The extent of the summit plateau so far to the northwest of the central part of the island must be explained in connection with this problem; the reduction of the area of the plateau apparently has taken place primarily from the northeast and southwest; the headward growth of the canyons on the northwestern side has been slow. Structural and topographic relations within the plateau area and in the sections immediately adjacent thereto are difficult to ascertain owing to climatic conditions and the abundant cover of vegetation.

Powers thinks that a second volcano was originally present on the northwest side of Kauai, and later was downfaulted below sea level but this is not supported by field evidence nor by a study of the submarine contours. (See pp. 52-53.)

The Emerged Wave-Cut Platform

The Hawaiian islands lie in the northern marginal belt of the coral seas where reef-building organisms exist with some difficulty and consequently do not develop luxuriant reefs. The reefs bordering the windward islands are feeble, and apparently are of late Pleistocene or Recent age. Conditions since have become so unfavorable that the reefs about Kauai and Niihau at least are now practically extinct. It seems probable that no reefs existed about any of the windward islands and possibly not about leeward islands in the earlier Pleistocene or in preceding periods, hence the islands have not been protected from wave attack by these structures to any extent; prominent abrasion features therefore are present along their shores.

For a long time after the close of the main volcanic period, Kauai remained relatively stable and was exposed to the powerful attacks of the open ocean waves. A broad platform was cut into the dome and from its inner margins great sea cliffs rose on all sides. Then diastrophic movements set in, and an intermittent tilting of the island brought the eastern half of the wave-cut plain above sea level so that the former shore line stood at an elevation of 250 to 500 feet. The western half was further submerged; the only records of its existence are from well borings located on the constructional plain of southwestern Kauai. A lava basement with overlying soil and weathered rock has been met in well borings at depths of 275 to 300 feet below sea level. This basement I believe to be the equivalent of the emerged marine platform on the eastern side of the island. On that side a large acreage was brought above sea level and the topography of the island was notably altered. Thus, at the base of the steep slopes of the eastern highland, which are the greatly battered remnants of former shore cliffs, is a rudely crescentic plain, in most places cliffed along its seaward margin (fig. 3; Pl. IV *B, C*). A number of tuff cones and residual basaltic ridges break the evenness of its surface. The platform as it emerged was a composite structure; wave-bevelled flows were exposed along its inner border; farther seaward, layers of basaltic detritus with which was intermingled a certain amount of organic material coated the rock basement. During the course of emergence, the detrital covering, except for certain boulder beaches, and smaller amounts of finer débris, was gradually swept seaward, and practically none of it now appears on the exposed portions of the plain. (See Pl. V, *B,* C.)

The platform is present only on the eastern side of the island between Hanalei Bay on the north and the mouth of the Waimea River on the south;

at the latter place it adjoins the constructional plain of southwestern Kauai. On the northern and southern sides, the platform varies in width from 1 to 4 miles; and on the eastern side, from 1 to 8 miles. The gradient is 60 to 100 feet per mile. The great width of the eastern section is due to the extensive down faulting which took place there prior to the erosion of the platform. The floor of the engulfed area was bevelled practically to the base of the southern, western, and northern scarps. Since emergence, marine abrasion has considerably reduced the area of the platform, especially along the northern and northeastern margins where cliffs 100 to 250 feet in height have been cut; along the eastern and southern margins, these recent cliffs rarely exceed 50 feet in height.

The time required for the erosion of such a platform as that of eastern Kauai is long and requires more or less continued stability of the island. Wave action very likely was virtually uninterrupted. Though the climate may have been suitable for the growth of reef-building organisms, reef probably did not develop during the early erosional stages because of the narrowness of the platform and the abundance of detritus continually rained on to it by the actively eroding youthful streams and by the waves which were attacking the lavas along the shore line. Even after the platform increased in width, there is no evidence that reefs surmounted it until comparatively recent times. Thus the shores appear to have remained unprotected and erosion actively proceeded; continuously steepened cliffs were present and the platform was gradually widened until the tilting of the island commenced.

PLATFORM IN THE WAILUA DRAINAGE BASIN

On the eastern side of Kauai, the development of the volcanic sink has materially affected the topography of the dome. (See figs. 6, 7.) A considerable engulfment took place, but its magnitude cannot be determined owing to the subsequent changes wrought by fluviatile and marine erosion. In this section, the widest portion of the marine platform was later developed, in spite of the fact that two ridges (Nonou and Kalepa) paralleling the eastern shore must have materially reduced the force of the waves. The walls of the sink are breached, so that the southern rim (Haupu Ridge) is now separated from the main highland; a narrow gap also divides the shore ridges, and these in turn are isolated from the north and south highlands by similar depressions (fig. 3). These gaps may represent breaches made by minor faults traversing the walls of the sink or by the enlargement of former drainage lines by wave action. The Nonou-Kalepa gap appears to have been of the latter origin, since there is no evidence of faulting associated with it. The origin of the other breaches could not be determined. Thus, when the island stood at the level at which the marine platform was developed, Haupu,

Nonou, and Kalepa ridges, together with a number of small erosion remnants, were islands in the pre-emergence sea, and the waves swept through the narrow straits between them (fig. 9).

The western and northern walls of the sink have been so greatly dissected by the numerous streams coming from the rainy highlands above that the original scarp topography and the sea cliffs at the base of the spurs have been largely obliterated. The rainfall over Haupu, Kalepa, and Nonou Ridge is comparatively light; hence these sections of the wall have been eroded by

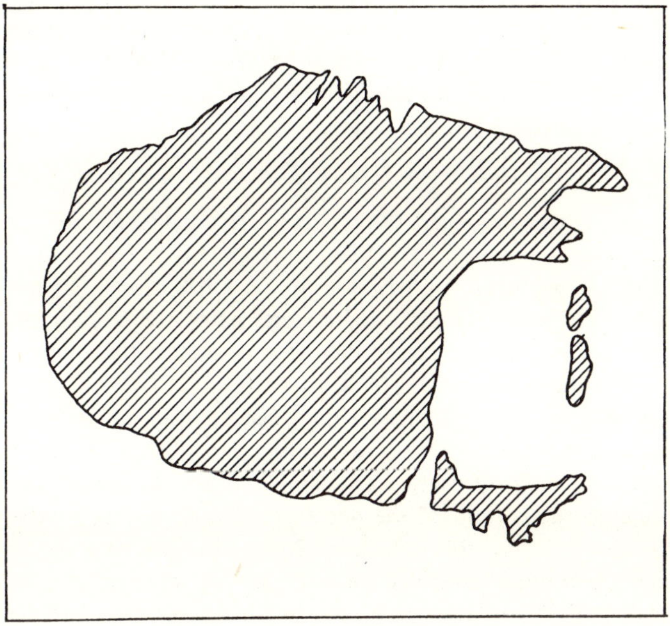

FIGURE 9.—Pre-emergence outlines of Kauai. Compare with FIGURE 3.

streams only to a slight extent. It is further probable that the steepness of the Haupu face was maintained by the sweep of waves through the Haupu-Kalepa gap.

The spurs between the valleys of the eastern (Waialeale) and northern (Makahela and Anahola) scarps exhibit less evidence of wave truncation than do those of the former northern and southern coasts of the island, apparently because of the protection afforded by the remnants of the sink wall paralleling the eastern coast and because of the dissection of any existing facets by numerous streams. The best preserved sea cliffs in this eastern section are present along the southern base of the Anahola Mountains; the wide gap between these mountains and Nonou Ridge seems to have allowed a more unbroken sweep of the waves against the ends of the spurs.

The date of faulting which resulted in the formation of the sink cannot be determined, but, judging from the extent of the erosion of certain of the scarps, it must have been early in the postvolcanic stage. The canyons which have been incised into the northern and western walls are comparable in depth with those of the deeply eroded northern and southern sections of the island, hence they have required approximately the same time for their development. The sink was formed prior to the erosion of at least most of the marine platform, since its floor was bevelled and the interfluves of the scarp valleys were truncated by wave action when the island stood at the level at which the marine platform was cut. Considering the length of time necessary for the erosion of the marine platform and for its emergence and subsequent dissection, the engulfment must have occured well back in the Pleistocene or even in the Pliocene.

POSTEMERGENCE DISSECTION OF THE PLATFORM

During and after emergence, the highland streams, which formerly flowed into the ocean at the mouths of their canyons, were extended across the platform and incised deep, narrow gorges into the newly exposed surface (Pl. IV, *C*). The presence of broad alluvial flats in the lower mile or two miles of these valley bottoms proves that the mouths of the valleys were eroded for some distance below the present sea level (Pl. V, *A, B*). Borings in Hanapepe Canyon 1¾ miles from the coast showed alluvial sediments 60 feet below the present sea level; excavations for bridge piers near the mouth of the Wailua River were carried into alluvium for 80 feet below sea level without reaching the bottom of the deposit. The deepening of the gorges I believe took place during the periods of lowered ocean level in the Pleistocene, though no positive evidence on this point is obtainable. During and after the rise of the oceans which accompanied the retreat of the continental ice sheets, the embayed valley mouths were gradually filled with alluvial deposits which finally rose above sea level except at the mouths of the valleys where wave action has been sufficiently strong to carry the sediment seaward and thus prevent the accumulation of deltal deposits. Of course, similar topographic features may be produced by the subsidence of an island, but, on Kauai, there is no evidence that such a subsidence has taken place. The form of the upper and lower sections of the platform valleys thus is strikingly different. The upper portions are narrow, V-shaped gorges; the lower portions are steep-walled, flat-bottomed valleys in which the streams meander slightly over the alluvial deposits. The mouths of all the larger platform and highland valleys are embayed. Between the valleys, broad, tabular, practically undissected remnants of the wave-cut surface appear.

A few isolated basaltic eminences, which stood as islands in the preemergence sea, and cinder cones of comparatively recent origin constitute the most conspicuous positive relief features on the platform (Pl. V, *B, C*). The larger of these residuals are the remnants of the south and east walls of the volcanic sink (Pl. V, *B*); their elevations and dimensions are as follows:

	ELEVATION (feet)	LENGTH (miles)	WIDTH (miles)
Haupu Ridge (south wall)	2,280	8	1-2.5
Kalepa Ridge (east wall)	710	3.75	0.75
Nonou Ridge (east wall)	1,328	2	1.25

Inside the walls are two basaltic eminences, and, near the east base of Waialeale, are two other small hills of about the same size. These two hills may be either cinder cones or residuals; their surface materials are so thoroughly altered as to prevent identification. On the northern side of the platform, Kamookoa Ridge is a spur which was isolated from the highland and stood as an island in the pre-emergence sea.

ORIGIN OF PLATFORM

The evidence proving that the coastal lowlands of the eastern half of Kauai have been formed by the emergence of a marine abrasion platform rather than by the work of streams may be summarized as follows: (1) the platform slopes from the base of the highland toward the shore line rather than from a series of divides toward the various river valleys; (2) the ends of the spurs along the inner margins of the platform are frequently sharply truncated, hence they mark the line of former shore cliffs; (3) the prominent ridges paralleling the shore of eastern Kauai have been wave cliffed; (4) the surface of the platform is more even and its gradient more uniform than if the platform were the product of fluviatile erosion; (5) at various localities, minor terraces have been cut into the main structure indicating apparent interruptions in the emergence; (6) boulder beaches are present at many places along the shoreward margin of the platform.

From observations of the marine plain and the two ridges paralleling the eastern shore made during "a rapid walk" about the island, Dana (19) expressed the tentative opinion that the plain derived its lavas from the volcanic mountain which subsequently arose within the limits of an older, partially destroyed volcanic mountain of which Kalepa and Nonou ridges would be remnants. An examination of the sections exposed in the gorges of this eastern plain showed that lavas erupted from the interior of the island underlie the flows exposed in the shore ridges and also underlie the flows of Haupu and Anahola ridges which project eastward from the interior highlands on the northern and southern margins of this plain. The plain in turn has been eroded into these lavas. The flows of the ridges are therefore younger than those whose weathered products form the surface of the plain, and evidently are part of a series once continuous with that visible in the great Waialeale scarp to the west and in the Anahola and Haupu ridges to the north and south. The shore ridges therefore cannot be interpreted as remnants of an older volcano but are parts of the present dome isolated when the engulfment took place.

Hitchcock's opinion (40, p. 13) that the eastern plain has been constructed by eruptions from a number of subordinate cones is not supported by field observations.

PARASITIC TUFF CONES

After the emergence and partial dissection of the platform, volcanic activity developed at a number of subordinate centers, located chiefly on the marine plain; small lava flows and tuff cones were erupted (Pls. V, *C;* VI, *A*). On the southern section of the lowland, where many tuff cones appeared, much of the pre-existing topography was obscured, and elsewhere smaller sections were buried beneath these relatively recent deposits. The location of the various cones is shown on the geological map (Pl. XIII). Most of the cones stand less than 300 feet above the surrounding surface and do not exceed 1½ miles in diameter. Kilohano, however, on the eastern plain about 4 miles west of Lihue, rises about 700 feet above the plain and has a diameter of nearly 5 miles (Pl. V, *C*). The slopes of this cone are gentle, not exceeding 5 degrees on any radius. The low gradients are due in part to the presence of a considerable number of lava flows interbedded with the ejectamenta. The principal cause, however, has been the heavy rainfall over the surface of the cone, which has made for active solifluction; thus the elevation and slope have been reduced and the diameter magnified.

The eruptions at the various vents did not take place simultaneously as is shown by the difference in the degree of weathering of the ejectamenta and in the stage of dissection of the cones in sections of the island having approximately the same rainfall. The materials of the cones of the southern lowland have been lateritized at the surface, and their original characters are difficult to determine. In the Kilohana cone on eastern Kauai, the rock structures are still visible, but the ejectamenta have been thoroughly altered. Mauna Ou and Hanahanapuni, on the eastern section of the platform, which possibly may be tuff cones, have had their materials weathered beyond recognition; these eminences possibly are residuals. On the northern side, the tuffs and breccias of the Kilauea cone are comparatively fresh; those of Kamoku and Ka Loko near the northern base of the Anahola Mountains compare in stage of alteration with the materials composing the cones of southern Kauai. The ejecta of a cone (Pohakea), near the outer margin of the southern highlands, compare with Kilauea in the stage of alteration. Thus Kilauea and Pohakea appear to be the youngest cones; the relative ages of the others could not be determined.

The outlines of the cones have been considerably modified by fluvial and marine abrasion. A few have their craters preserved, but most have been reduced to rounded domes. The extent of fluvial dissection is a factor of the rainfall over the area in which the cone is located and of the variations in

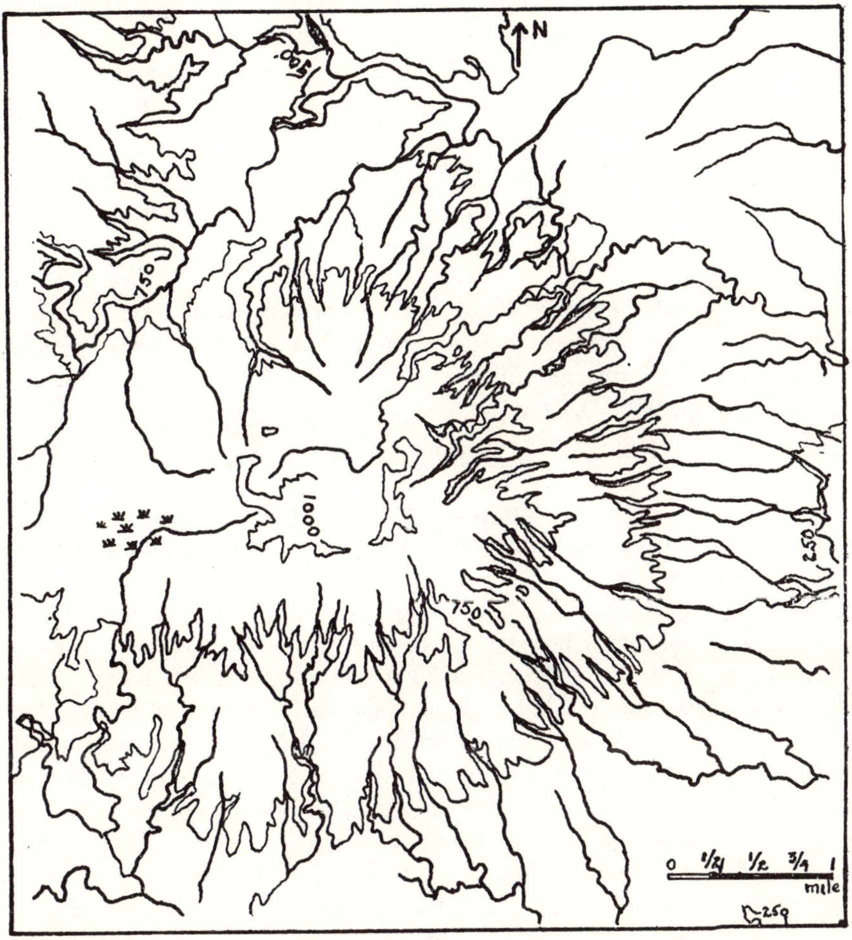

FIGURE 10.—Topographic map of Kilohano tuff cone, eastern Kauai. The asymmetric development of streams and valleys on the windward and leeward sides is illustrated. Contour interval, 250 feet.

rainfall over different parts of the cone itself. In Kilohana in particular, the stage of erosion on the windward and leeward slopes is strikingly contrasted—a reflection of sharp differences in rainfall within the narrow limits of the diameter of the cone (less than 5 miles). Unfortunately there are no quantitative measures of precipitation over various parts of the cone; at the single station located at the summit, the annual average is 60.15 inches. Without question, the heaviest rains fall over the windward slopes. The northern and eastern sides are cut by many deep valleys, between which are high narrow ridges. On the southern and western sides, the valleys are shallower, wider, and less numerous, and are separated by broad, conic interfluves (fig. 10). The vegetation presents further evidence of the difference in precipitation, since the windward slopes are covered with a rather dense growth of trees, ferns, grasses, and brush of various sorts. On the windward slopes, trees are widely scattered, and the divides and large sections of the valley walls are covered only with grass. The trees, brush, and ferns are largely confined to the valley bottoms. Most of the permanent streams are on the windward side. Other cones show similar features, though less conspicuously developed.

More than half of the small cone on the northern coast, west of Kilauea Bay, has been destroyed by marine abrasion. A detached remnant of the tuffs and breccias forms a small sea stack just off shore from the central portion of the cone (Pl. VI, *A*).

Southwestern Constructional Plain

Of different origin and topographically quite unlike the marine platform is the crescentic lowland (Mana Flats) which adjoins the platform at the Waimea River and extends along the southwestern and western sides of the island as far as the Barking Sands (fig. 3). As previously stated, the tilting of the island which brought the eastern half of the wave-cut platform above sea level further submerged the western part so that it now stands about 300 feet below sea level. Upon this platform grew a fringing reef which was later killed and covered by débris eroded from it and from the basaltic highland to the east. Part of this deposit now stands above sea level as a coastal plain which abuts against the stupendous, sea-cliffed face of the western highland. The greatest width of the plain is about 2 miles; nowhere does it exceed 50 feet in elevation above sea level. Its surface is comparatively flat. A coastal rampart of sand dunes, rarely more than 30 feet in height, constitutes the principal relief. Shallow, salt marshy depressions of considerable size are present at a number of places behind the coastal dune belt. Along the base of the highland is a more or less continuous steeply sloping apron of talus and alluvium which not infrequently rises a hundred or so feet above the surface of the plain. The illustrations of the Haena reef plain of northern Kauai (Pls. VI, *B*; VII, *A*; IX, *A*) are typical of the relief of the Mana flats. No valleys cross these lowlands. The flood waters which flow through the canyons of the highlands during the heavy rains are quickly absorbed in the porous soil and rarely reach the sea.

Other Constructional Plains

Similar plains of smaller extent, formed over and behind the reef fringing northern Kauai, are found between Haena and the mouth of the Wainiha River, east and west of the Kalihiwai River, and east of the Kilauea River. In these three areas, the reef extends for some distance beyond the seaward margin of the plains; lagoons are generally present over the unburied reef surface (Pls. VI, *B*; VII, *A*, *B*; IX, *A*). Along southwestern Kauai, the limestone has been completely buried, so that the subsurface structure cannot actually be demonstrated. Its presence below the surficial deposits I think is highly probable. Another mode of accumulation of the débris in the latter area must, however, be recognized. Talus and alluvial material may have piled up in considerable thickness on the sinking platform at the base of the cliffs. Calcareous sand drifted by the strong currents running along the northern and southern sides of the island also may have found lodgment in this area. The resulting deposit gradually grew up to or near sea level, and then was exposed by the recent slight change of level which has affected the island.

COASTAL TOPOGRAPHY OF KAUAI

The shore line pattern of Kauai is relatively simple; the principal embayments are the submerged mouths of the larger streams crossing the marine platform of the eastern half of the island. These valleys appear to have been deepened during the lowered ocean levels of the Pleistocene and then to have been drowned as the sea returned to its present position. Hanalei Bay on the northern coast and Nawiliwili Harbor on the eastern coast are the principal indentations.

Before the emergence of the marine platform, the island had been strongly cliffed on all sides. The cliffs of the windward side were dissected by the numerous streams, and were further battered after the exposure of the platform ended marine erosion. These former sea cliffs are now greatly eroded high angle slopes, which show only to a limited extent the evidence of their origin. The stupendous cliffs of western Kauai, 500 to 2,000 feet high (Pls. VIII, *A*, *B*), suggest that they were cut when the prevailing wind direction in this region was different from that of today, as it may well have been during portions of the Glacial Period. A narrowing of the tropical belt as the chilling of the earth's climate progressed may have brought the stormy westerlies as far south as the latitudes of Hawaii, so that the principal wave attack for a time was along the western coast. Daly (17, p. 92) has suggested a similar shifting of the wind belts in the Atlantic during the Pleistocene to account for the giant cliffs of leeward St. Helena.

Along the northwestern and southeastern coasts of Kauai are a few sea stacks. Into the base of the cliffs, many sea caves have been eroded, and through narrow headlands tunnels have been cut (Pl. VIII, *B*).

The growth of the now extinct fringing coral reef upon the submerged platform of western and southwestern Kauai has protected the cliffs of that section from further wave abrasion. Between the Barking Sands and Haena on the northern coast, only a few patches of reef are present.

A notable feature of the shore pattern of Kauai is the straightness of the northwest coast; this has been explained by Dana (19, 21) and by Powers (52) as the result of down faulting. While I agree with this opinion, I am convinced that the down faulted mass was comparatively small and was not a second great dome as Dana and Powers believe (pp. 52-53). If faulting did take place in this locality, it has not been of recent date, hence the cliffs have retreated for some distance from the zone of dislocation. The width of the wave-cut bench at the base of the cliffs is not evident from the hydrographic charts. The northwestern cliffs are the highest on Kauai and compare in majesty with those on any of the other islands (Pl. VIII, *A*); they have been greatly dissected by streams, and have been recently nipped by invigorated wave attack at their base. The new cliffs average about 250 feet in height. Along the shore line are sea stacks, many great sea caves, tunnels through narrow headlands, and a few patches of extinct reef or small colonies of reef-building organisms (Pls. VI, *A, C;* VII, *B;* VIII, *B*).

The great spur of the eastern highlands which projects to the southeastern coast (Haupu Ridge) also is sharply cliffed, and has suffered recent nipping.

Since the emergence of the marine platform, sea cliffs have been cut into its shoreward margin; these vary from 250 feet in height along the northern and northeastern coast and decrease to less than 50 feet along the eastern and southern coasts (Pls. VI, *B*; VII, *C*). These cliffs in places are protected by patches of the Recent fringing reef.

The southwestern plain slopes from the base of the highlands to the shore line except where the low rampart of sand dunes is present along the coast. On the northern side, the constructional hinter-reef plains slope in similar fashion, but the fringing reef projects for some distance beyond their seaward margins (Pls. VI, *C;* VII, *A, B;* IX, *A*).

A recent slight lowering of sea level has brought narrow, wave-cut terraces 5 or 10 feet above sea level in many places; these terraces rarely exceed a hundred or so feet in width.

DRAINAGE

DRAINAGE BASINS

According to Martin and Pierce (45, p. 36), the drainage basins of Kauai (fig. 2) are determined by three great ridges which branch off from Waialeale and Kawaikini, the central and highest points of the island:

> One of these ridges extends toward the northeast through peaks back of Kealia and along the crest of the Anahola Mountains to the sea. The other two constitute the main divide or backbone of the island, which, starting at the southeast, follows along the Haupu Ridge east of Koloa across the Koloa-Lihue Gap and then northward along the ridge east of Hanapepe basin to the summit, where it turns slightly to the northeast along the western edge of the Wainiha Basin to the sea. Another important divide leaves the main one at Kilohana, north of Alakai Swamp, and follows westward along Kaunuohua Ridge, and then southward along the western edge of Waimea Canyon to the sea. These watersheds mark out four distinct drainage areas or basins.

On the basis of this interpretation, a number of streams flowing from the summit plateau (Alakai Swanp) to the sea along the northwestern coast are included in the morphologically unrelated western drainage basin where the streams are fed by the scanty precipitation falling over the region to the west of Waimea Canyon and not from the waters of the Alakai watershed. A more natural bounding of the northern basin is as follows: The western rim of Wainiha Canyon from Waialeale to Kilohana, from the latter point westward along Kaunuohua Ridge, and then northwestward to the sea along the divide between Awaawapuhi and Nualolo valleys. The Awaawapuhi is the westernmost stream deriving its principal water supply from Kaunuohua Ridge, a part of the summit plateau.

For convenience I have named three of the drainage basins from the principal streams running through them—the northern or Hanalei Basin, the eastern or Wailua Basin, the southern or Hanapepe-Waimea Basin. As there are no dominant drainage lines in the western basin, I have applied to it the name "Mana," from the local name of the plain bordering the highland on the west. The boundaries of these basins are shown in figure 2. The Hanalei Basin receives on the average the heaviest rainfall of the island, hence its streams are the largest and most numerous. The chief drainage lines are the Wainiha, the Hanalei, and the Lumahai, while of less importance are the Waioli, the Kalihiwai, the Kilauea, and the Moloaa. The streams crossing the section of the basin west of the Wainiha River are smaller but their short canyons compare in depth with those farther to the east. The source of the drainage flowing through this basin is either in the bogs of the summit plateau or from precipitation which falls over the crests and northern slopes of the Anahola Mountains, the great ridge forming the southeastern boundary of the basin. As with all of the streams of Kauai and of other parts of Hawaii, underground drainage, which migrates rapidly through the porous and cavernous lavas, is an important or even a principal source of the surface streams. The main drainage lines of the eastern or Wailua Basin are the two forks of the Wailua River, which flow from the summit plateau and the adjacent mountain slopes. There are several minor streams which drain areas to the north and to the south of the Wailua system. The largest rivers of the southern basin are the Hanapepe, the Olokele, Mokihana and the Waimea. Through this area passes most of the run-off from Alakai watershed. The western, or Mana Basin, is nearly cut off from the Alakai watershed by the aberrant, southward flowing Waimea River. Because the precipitation over the basin is light, and occurs very largely during the rainy season, most of the streams have water in them only during the heavy rains. The valleys of this section therefore are sharply contrasted in form with those which have been produced by the powerful streams of the other sections of Kauai. (See fig. 8, *b*.)

In general, the courses of the streams flowing from the summit plateau are roughly radial; the north-south channel of the Waimea River, one of the principal streams of the island, is, however, an important exception. A fault line apparently has deflected the drainage lines (the Koaie, the Waialae, the Oomau, and the Mokihana), which normally should have flowed westward and southwestward from the summit plateau, and made them tributary to a master channel following the course of the dislocation. The Waimea and the Makaweli-Olokele unite near the sea. The Olokele is a southwestward flowing stream, whose course appears to have been interrupted by the Waimea fault not far from the shore line; consequently only a slight deflection of its direction has taken place (fig. 8, *a*).

FLOW OF STREAMS

The streams of Kauai are numerous and large, and, according to Martin and Pierce (44) they are more uniform in flow than are those of the other islands. Nevertheless, they are subject to considerable fluctuations in volume as the result of the marked seasonal distribution of rainfall. (See Table 7.) West of the Waimea, the streams are short, intermittent, and have few tributaries. The west rim of Waimea Canyon is the divide which separates the drainage of the Mana Basin from that flowing southwestward from the Alakai Swamp. The drainage map (fig. 3) also shows the great contrast in the complexity of the drainage patterns on the windward and the leeward sides of Kauai.

An average of approximately 80 second-feet a month of the waters of the Waimea River is diverted for irrigation purposes above the station at which the discharge measurements are made. During the dry season, the water so used is a very large proportion of the total flow of the stream.

Diversions of the water from Hanapepe River for irrigation purposes average about 40 second-feet per month.

While in general the minimum discharge of the various streams is recorded during the dry season, there are numerous exceptions to this statement. This applies especially to the streams having their sources in the boggy summit plateau or on the adjacent ridges where the rainfall throughout the year is more evenly distributed. (See Table 2.)

TABLE 7. MONTHLY DISCHARGE OF KAUAI RIVERS (MEASURED IN SECOND-FEET)[a]

	Waimea River				Hanapepe River				Hanalei River			
	1913	1914	1915	1916	1913	1914	1915	1916	1913	1914	1915	1916
July	117.0	38.5	8.5	49.8	45.2	87.3	49.5	No	184.0	190.0	82.3	235.0
August	16.9	42.2	17.3	25.2	27.4	123.0	77.7	Record	136.0	266.0	99.8	183.0
September	52.0	294.0	46.1	14.0	18.7	317.0	71.6		116.0	1050.0	111.0	214.0
October	254.0	63.0	72.7	107.0	57.0	93.6	127.0		112.0	325.0	134.0	313.0
November	1640.0	47.5	480.0	238.0	166.0	115.0	271.0		224.0	323.0	297.0	486.0
December	37.7	382.0	347.0	534.0	27.1	206.0	126.0		12.4	610.0	162.0	636.0
January	183.0	19.5	1640.0	367.0	22.1	17.5	196.0		112.0	140.0	275.0	535.0
February	81.1	76.4	209.0	299.0	13.8	24.8	93.1		91.0	153.0	84.6	164.0
March	232.0	1.78	320.0	514.0	19.8	15.9	114.0		137.0	93.3	158.0	956.0
April	905.0	179.0	105.0	133.0	30.8	110.0	34.3		237.0	424.0	52.3	210.0
May	62.5	17.5	184.0	183.0	92.4	22.4	96.0		320.0	203.0	136.0	441.0
June	39.0	56.8	50.4	91.6	109.0	56.5	80.9		330.0	266.0	88.7	398.0
Yearly	229.0	101.0	291.0	214.0	52.5	99.2	112.0	....	176.0	337.0	322.0	401.0

[a] Data compiled by Grover (26, 27), Grover and Larrison (28).

SURFACE FLOW AND UNDERGROUND DRAINAGE

In Hawaii, the most casual observations show that the amounts of run-off and of water penetrating below the surface vary greatly at different times in the same area and vary permanently over different sections of the same mountain. This is notably so on the higher, better watered domes of the group.

On Kauai, no quantitative measurements of the relation between run-off, evaporation, and accessions to the groundwater supply have been made, but certain general facts are perhaps worthy of brief consideration.

1. On the mountains of Kauai, the windward slopes receive much heavier precipitation than do the leeward, (fig. 2). Because of the more abundant run-off and groundwater, the windward rivers are larger, more permanent, and less subject to great fluctuations in volume.

2. The storms vary greatly in intensity; during the rainy season, the fall of water is extremely heavy, so that very high maxima are commonly measured for brief periods of time. The precipitation over the whole island is in the form of rain. It is evident that the accessions to the groundwater

supply are much more per unit of water falling during the light rains of the drier months than during the torrential rains.

3. According to existing records, Kauai is the most heavily watered of the Hawaiian islands; the greatest annual average measured elsewhere in Hawaii is 350 on the summit of West Maui. The areas on Kauai over which average annual rainfall of 80 inches or more fall are as follows:

INCHES	SQUARE MILES (approximate)	INCHES	SQUARE MILES (approximate)
450	10	250	50
400	10	200	55
350	15	150	110
300	30	100	120
		Under 80	135

Over the Kauai lowlands, most of the rain falls during the winter months; on the highland slopes, it is much more uniformly distributed throughout the year.

4. The relief of most parts of Kauai is extremely rugged, and the slopes are steep. The elevated marine platform, the various constructional plains, and the summit plateau have gentle slopes. The constructional plains have the least diverse surfaces and the lowest slopes.

5. Except in the rainy sections of the island, the soils are porous and allow the ready penetration of groundwaters. Even through the more compacted clayey soils of the heavily watered upland, great quantities seep. The most porous surficial materials are the moderately coarse grained calcareous and basaltic sands of the constructional lowlands, and through these even the torrential flow from the highland valleys quickly sinks. Below the soil cover and the zones of deeply weathered rocks, the fresher lavas are abundantly supplied with openings; vesicles; vesicular, scoriaceous, and brecciated flow tops; joints of all sizes; cavernous spaces between the various layers of many flows and between the flows themselves; and faults. Even though much compacting has taken place from the crushing weight of the overlying lavas, the rocks exposed farthest below the surface have innumerable channels through which waters can penetrate. A few extremely dense flows and many of the sills and narrow dikes are relatively impervious.

6. Except in the rainiest sections, the saturation of the soils and immediately underlying rocks varies with the seasons. During the rainy months, the water table is close to or at the surface, but falls rather rapidly during the drier portion of the year. In the uplands, the more or less continuous fall of rain keeps the soils well saturated. Farther below the surface, certain of the porous zones evidently carry nearly their maximum quantity of water, but most are far from saturated.

7. The lavas dip seaward from the central part of the island at angles generally less than 10°. This quaquaversal structure, combined with the

high porosity of the rocks, makes for the active migration of the groundwater solutions. Downward penetration and migration along the dip are rapid. The meteoric supply works its way for some distance below sea level, and has sufficient head to force back the ocean waters which otherwise would occupy the openings. Wells drilled far enough from the sea and reaching zones in which the flow of water is rapid remain fresh. Migration of groundwaters and the abundance of the supply are apparent along most of the mountain slopes; many streams are fed from spring lines and not from the surface flow. Not infrequently these springs are located hundreds of feet below the crests of the interfluves. According to Mr. L. D. Larsen, Manager of the Kilauea Plantation, a considerable part of the drainage from the northern slopes of the Anahola Range comes from springs which emerge at an elevation of about 1100 feet, which is in places over a thousand feet below the top of the mountains.

8. The vegetation cover in the rainy sections plays an important part in checking the rate of run-off from the steep slopes. Over the lowlands and the leeward sections the plant cover is relatively scanty.

KAUAI-NIIHAU VOLCANO

A study of the hydrographic charts of Kauai and Niihau shows that these two islands are the summits of a compound volcano, one of the mountains in the Hawaiian range, which extends from an unnamed shoal 280 miles west-northwest of Ocean Island 1900 miles southeastward to the island of Hawaii. On the basis of existing data, the submarine contours cannot be drawn with sufficient detail to indicate whether or not submerged major eruptive centers exist, though there are certain indications that such is the case. As on Oahu, Maui, and Molokai, the western member (Niihau) of the pair of domes first became extinct. Whether Kauai and Niihau ever were united above sea level cannot be determined. Since downfaulting has caused the disappearance of the eastern half of Niihau, the islands were once more nearly contiguous than at present, but I do not think that exposed connections ever existed between them. The original relations probably were similar to those now existing between Maui, Molokai, Lanai, and Kahoolawe—these four sections of a single huge volcano are separated from each other by narrow channels of relatively shallow depths.

The submarine connection between Kauai and Niihau is shown by map and cross section (Pl. X; figs. 11, 12). The contours have been plotted from

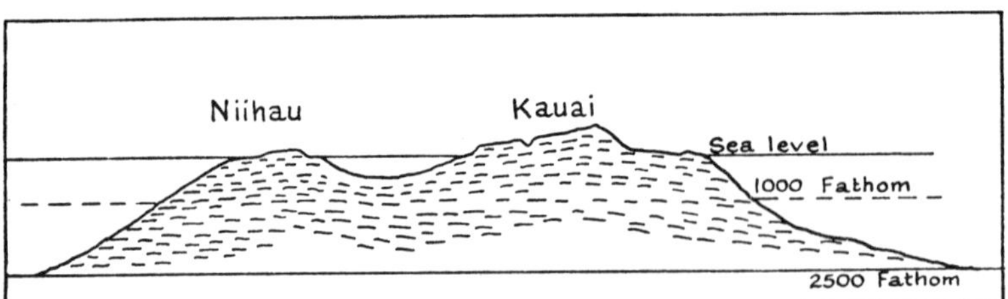

FIGURE 11.—Cross-section of the Kauai-Niihau volcano. Horizontal scale: 1: 1,950,000; vertical scale: 1/10 inch=5000 feet.

data given on Hydrographic Chart 4117, published by the United States Coast and Geodetic Survey, and extend to depths of 1000 fathoms; beyond this limit, only a few scattered soundings have been made. While the floor of the ocean about the mountain averages nearly 2,500 fathoms in depth, it is probable that the form of the volcano does not change greatly below the 1000-fathom line. The soundings at lesser depths are too few to permit an exact delineation of the submarine relief, but the general relations are evident. Soundings off shores of Niihau are fewer than off Kauai, hence the contouring of the submerged slopes of that mountain is incomplete, especially on the southern side. It is impossible, for example, to determine the relation of

Niihau to the small tuff cone, Kaula, which lies 19 miles to the southwest. Kaula certainly cannot be a parasitic tuff cone on the submerged flank of Niihau, since, if the slope of that dome be more or less uniform, Kaula would rise from a depth of over 2,000 fathoms. It is reasonable to postulate that the cone rises from a third dome whose presence has not been recognized because of the scarcity of the soundings. If the summit of this dome approached fairly close to sea level, the height and mass of Kaula would not be very different from similar dimensions of other cones in the archipelago.

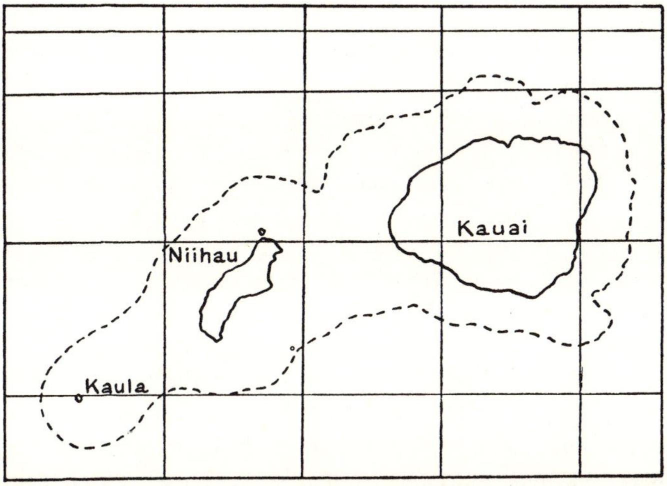

FIGURE 12.—Ground plan of the Kauai-Niihau volcano at a depth of 1,000 fathoms. The submarine connection of the submerged platform upon which the tuff cone, Kaula, is imposed is shown by the dotted line. Scale: 1: 1,800,000.

Kaula has been described in some detail by Palmer (48, pp. 6-10), and a survey of the island and surrounding waters was made recently by Captain C. L. Garner of the U. S. Coast and Geodetic Survey. The survey shows that the tuff cone surmounts the eastern edge of a submarine platform which has an area of approximately 30 square miles. The depth of water over the platform varies from 27 to 53 fathoms, while away from its margins, the depths exceed 200 fathoms. At a point about 3 miles northwest of Kaula, a rocky pinnacle extends to within 38 feet of the ocean surface. This survey confirms the view that Kaula was imposed upon a submerged dome. This mountain very likely once rose above sea level, and has long since been destroyed by erosion. Palmer has found that a reef grew up on the submerged platform, and that through this the materials of the cone were erupted. The mountain represented by the submerged platform undoubtedly is a part of the Kauai-Niihau volcano.

About one-half mile north of Niihau is another tuff cone, Lehua (elevation 738 feet), which has been erupted on the northern slope of the dome.

From a brief reconnaissance on Kauai, Powers (52) concluded that the island

> . . . was originally a volcanic doublet, the western half of which has disappeared. . . . Dana was the first to notice that the flows truncated in the Na Pali cliffs dip toward Waialeale; but he supposed that Niihau represented the missing volcano, moved southward. . . . Niihau is but a remnant of the original volcano, the center of which was east of the present island . . . . The northeastern part (of Niihau) is bounded by a fault-scarp, along which the greater part of the original island has subsided; but this scarp is not a continuation of that at Na Pali.

I have not been able to find in Dana's papers (19, 20), a statement that the lavas of the northwestern section of the island dip toward the center at Waileale. Dana suggests that Kauai was originally a doublet island through which a great fracture zone developed having a trend approximately at right angles to the fissure system over which the archipelago has been built. Along this transverse zone, both vertical and horizontal movement took place. The vertical movement caused the disappearance below the sea of most of the "lost" dome; the horizontal caused the transference to its present position of the residual mass (Niihau).

I have made a detailed examination of the structural relations and of the sections exposed in the cliffs in a number of the canyons which opened to the sea along the Na Pali coast between Wainiha and Kolalau; in Honopu, Nualolo, and Milolii canyons to the west of Kolalau; and in certain of the smaller gorges opening onto the coastal lowland near the Barking Sands. In each of these canyons, except for unimportant local variations, the flows dip seaward at angles generally less than 10°, and thus correspond in their relations to those of the rest of Kauai. No general dips toward the central part of the island were noted. As Plate X indicates, the submarine contours do not show a great scarp off northwestern Kauai which could be interpreted as a zone of faulting along which a second dome had disappeared. The view held by Dana and Powers that Kauai originally was a doublet island is impossible of acceptance; the island always has been a single dome. Niihau is the remnant of a larger volcano, the eastern portion of which apparently has disappeared as the result of engulfment. The present cliffed face does not represent the line along which the dislocation took place, since the powerful wave attack along the coast has undoubtedly caused a considerable retreat of the scarp. The present cliffs of eastern Niihau therefore are a fault line scarp and not a fault scarp. The extent of the retreat from the fault zone cannot be definitely ascertained because of the few soundings which have been off eastern Niihau. Two depths of 40 and 42 fathoms respectively, 1½ to 2 miles off the southern end of the island indicate the presence of a submerged bench. A sharp change of slope is shown by soundings 250 to 400 fathoms within a half mile of the localities first mentioned. This bench may be continuous along eastern Niihau, but no depths have been measured close enough to the cliffed coast to prove its presence. The break in slope may represent the position of the fault scarp, and the bench the distance to which the cliffs have retreated under wave attack subsequent to the dislocation.

My field studies have confirmed Powers' observation that the lavas exposed in the cliffs of eastern Niihau have no relation to those of northwestern Kauai; hence Niihau does not represent a fragment of Kauai.

Washington's statement (59, p. 337) that Niihau is tectonically related to the leeward Hawaiian islands is out of accord with the evidence at hand. Niihau is separated from Nihoa, nearest of the leeward islands, by 140 miles of ocean averaging 1,500 to 1,800 fathoms in depth.. In geology, morphology, and structure, Niihau is related to the younger, windward section of the archipelago. Its close association with Kauai cannot be disputed.

IGNEOUS GEOLOGY OF KAUAI

STRUCTURE OF KAUAI DOME

The main igneous complex of Kauai is composed for the most part of thin, trappean or porphyritic flows of basalt and pyroxene andesite which dip seaward at angles of less than 10° from a principal eruptive area located near the center of the island. The great mass of the lavas has come from vents within the jungle-covered portions of the rainy highlands, hence their sources cannot be located; other flows have been erupted from central vents or from fissures located at various places along temporary constructional surfaces. Many of the feeders of this second group of flows are visible in the sections exposed in the great canyons of the drier parts of the island. The flows average less than 20 feet in thickness; very few exceed 50 feet. They gradually thicken toward their source. Their lengths and widths are difficult to measure because of the great amount of weathering and erosion which has taken place and because of the heavy cover of vegetation which hides much of their upstream sections. The lava is principally of the clinker (aa) type; the proportion of clinker to ropy (pahoehoe) surfaced flows is about three to one. Relatively few flows of blocky lava were observed.

Interbedded with the flows are occasional pyroclastic deposits but the volume of these is so small that the form and structure of the dome have not been affected to any extent by their presence. The deposits doubtless represent tuff and cinder cones erupted on temporary constructional surfaces which later were buried by lava floods. So far as can be determined, all of the ejecta are of basaltic or andesitic composition.

The main complex is cut by great numbers of trappean and porphyritic dikes which for the most part radiate from the central portion of the dome; certain groups, however, are associated with minor eruptive centers. The relations of the dikes are splendidly shown in the lower portions of the canyon walls and in the sea cliffs, but are completely obscured in the rainy highlands because of the deep regolith and the heavy cover of vegetation. The dikes rarely exceed 10 feet in width, and generally have been formed by single injections; a few multiple and composite dikes were observed. Great numbers of sills also are present; most of them are not more than 8 or 10 feet thick, but occasionally sill-like or laccolithic bodies up to 250 feet thick were found. Magnificent columnar jointing has developed in many of the dikes and sills and in some flows (Pl. II, *B*). Variations in the normal injection relations in both types of intrusives were noted; the dikes not infrequently are concordant or nearly so at one or more places in their courses and then assume their usual positions. The sills in like manner transgress from one horizon to another. In a number of places, the dikes are clearly flow feeders. A few, irregular, stock-like or chonolithic bodies were observed, but were so inaccessible that their relations could not be studied in detail. Boulders of coarsely granular rock found along a number of the stream beds indicate the presence of course-grained parent intrusives, but none was discovered. The location of some of the boulders suggests that they may have come from lava tube fillings or from sill-like injections similar to the Uwekahuna mass at Kilauea (57), though the texture of the rocks is distinctly coarser than that from Uwekahuna. These boulders do not appear to have been derived from the normal sills or dikes, neither are they to be associated with the pyroclastic deposits.

LAST VOLCANIC EPISODE

After the erosion of the dome practically to its present form, a later period of volcanic activity developed at a number of subordinate centers, most of which were located on the marine platform of the eastern half of the island; a few vents were opened in the highlands. Over a score of tuff cones and small lava flows were erupted. Along the southern lowlands, where activity was most intense, the topography of the platform has been considerably changed by the development of numerous tuff cones; some of these show distinct linear arrangement probably along fissures. The most extensive outwellings of lava came from fissures located near the base of Haupu Ridge, southeastern Kauai. This volcanic episode occurred in late Pleistocene or early Recent time, but sufficiently long ago to allow the deep weathering of the pyroclastics and the extensive erosion of most of the cones. The cones vary considerably in age, since, in areas of about the same rainfall, the degree of weathering and erosion is quite different. The position of the various cones is shown in Plate XIII.

PRINCIPAL ERUPTIVE CENTER

On Kauai, the principal eruptive area is hidden below the jungle tapestry of the rainy highlands. In such sections as are available, distinct flow structure can be traced on all sides close to the center of the island, but there is a considerable area in which no definite field data can be secured. In the central part, the flows are nearly horizontal or have an outward dip of not more than 3°; farther seaward the dip increases to 8° or 10°. The eruptive center or area thus appears to have been located in the region of horizontal or low dipping flows, but its nature could not be determined. If a sink were originally present, it has since disappeared; possibly it may have been largely filled with pyroclastics of which there are considerable quantities in the central part of Kauai. Fissure eruptions were numerous as the abundant dikes testify, but evidence was not secured to determine whether the island is of the Ascenion-St. Helena or of the Mauna Loa-Kilauea type.

No great central monolith, such as Dana (20) described from Tahiti is present on Kauai, since structureless rocks are not to be found in any of the sections visible in the upper courses of the canyons, nor are coarsely granular rocks of types which would form such a body present in the stream beds. If a central vent existed, the throat must have been narrow for some distance below the surface, and the lava which solidified therein relatively fine-grained.

IGNEOUS STRATIGRAPHY

At only one horizon in the exposed portion of the main igneous complex of Kauai is there evidence of a notable break in the eruptive sequence. In several of the canyons of western Kauai at elevations of about 500 feet, beds of course volcanic conglomerate overlie an erosion surface which has been cut into the lavas. The best exposures are in the bottom of Waimea Canyon, between the Waialae and Poomau streams. Unfortunately, the exposures are limited and little can be seen of the underlying erosion surface. The deposits are of sedimentary rather than of igneous origin, but whether they be marine or fluvial could not be determined. In the other canyons, this conglomerate was not observed, but conditions of erosion, the depth of rock weathering, and the heavy cover of vegetation may have concealed it. Whether this apparent break represents an important erosion interval is difficult to say, but I am inclined to believe that it does, and shall use the horizon to separate an older and a younger series of lavas composing the main complex. To the three groups of rocks, I shall apply the following names: Lower Kauai lavas, Waimea conglomerate, and Upper Kauai lavas.

Most of the older Kauai lavas were erupted below sea level, and the visible exposures are limited; the rock types are similar to those composing the upper series. The later series includes all of the intrusive and extrusive rocks erupted after the deposition of the Waimea conglomerate up to the close of the last principal eruptive cycle. This series apparently was not broken by any long erosion interval. Most of the flows have remarkably fresh surfaces, though a considerable amount of weathering has been accomplished by underground waters. Occasionally sufficient time elapsed between eruptions for the development of thin lateritic soils, but little erosion seems to have taken place. The lateritized zones are not continuous horizontally for any great distance, hence they are not of sufficient importance to be recognized as breaks in the stratigraphic column. The maximum thickness of the later Kauai series is about 4,000 feet. The thickness of the Waimea conglomerate probably does not exceed 75 or 100 feet, while that of the lower Kauai lavas cannot be determined.

After the close of the late Kauai volcanicity, minor eruptions took place at various centers, as they did on all of the Hawaiian domes. Most of these cones have been almost completely destroyed by erosion, and the pyroclastics remaining have been very deeply weathered.

The last phase of the eruptive cycle developed long after the sealing of the principal eruptive areas or vents and after most of the present erosional topography had been evolved. Many fissures and vents were opened on the emerged marine platform and a few in the highlands. More than a score of tuff cones were erupted, and a number of small lava flows buried various areas of the older lateritized rocks below. This activity developed after the emergence and erosion of the platform as is shown by the burial of parts of valleys which had been cut after the emergence. The eruptions took place in late Pleistocene or early Recent times, but were not contemporaneous. The pyroclastics have been weathered to different degrees, even in areas where there is little difference in the rainfall; some of them have been altered almost beyond recognition. Since most of these late outbursts took place in the Koloa District, southern Kauai, I have called the episode by that name.

The principal events of the igneous history of Kauai therefore are:

1. Eruption of the older Kauai lavas.
2. Erosion interval and deposition of the Waimea conglomerate. Whether this was a time of complete cessation of volcanic activity is not known.
3. Eruption of the younger Kauai lavas.
4. Erosion interval and emergence of marine platform. Minor eruptions appear to have continued from subordinate vents.
5. The late Koloa volcanic episode. Lavas and pyroclastics erupted from vents located mainly on the emerged marine platform.

IGNEOUS ROCKS

PREVIOUS INVESTIGATIONS

Details of the petrography of the Kauai lavas are even more meager than observations regarding the general geology of the island. Dana (19) noted the predominance of olivine basalt in the main igneous complex and the presence of a much later series of fragmental deposits and lava flows near Koloa on the southeastern coast which he described in some detail. In flows exposed in Hanapepe Canyon, Dana observed "crystals" of olivine more than one inch in length. Later studies have shown these to be small dunite xenoliths and not phenocrystic individuals of olivine. In a collection made by the zoologist, Schauinsland, in 1898, Kauai lavas are represented by three specimens from a locality given as Tipukai, which undoubtedly is Kipukai on the southeastern shore. According to descriptions later published by Möhle (47, p. 89), these rocks are porphyritic and aphanitic olivine basalts. Lindgren (43, p. 15) mentions a "coarse-grained olivine diabase" which was said to have come from "a large boulder in one of the principal streams of Kauai."

This rock was studied in the field and described by Cross, who collected from a number of localities on Kauai. Cross (7, p. 9) described two olivine basalts from Olokele Canyon; plagioclase basalt from a flow of the late volcanic episode near Kilauea; limburgite, containing inclusions of harzburgite and dunite, from boulders in the bed of Hanapepe Canyon near the falls; oligoclase gabbro (kauaiite), the coarsely granular "diabase" of Lindgren, from boulders of unknown origin near the mouth of Waialae River; and nephelite-melilite basalt, incorrectly associated with a late cinder cone near Kilauea on the north shore of the island. Analyses of the two olivine basalts, the oligoclase gabbro, and the nephelite-melilite basalt are given. Cross's report contains the only detailed descriptions of igneous rocks from Kauai.

Powers (53) made extensive collections from both Kauai and Niihau, but his report pays scant attention to the microscopy of the lavas. He emphasizes the abundance of olivine basalts and notes the presence of the late series of lavas and pyroclastics (the Koloa series). He describes briefly dunite nodules in a flow of picritic or limburgitic basalt near Puu Kahoaea, west of Lihue, and records other localities where similar inclusions are present. The mineral composition of a gabbro xenolith from a flow near the falls in Hanalei Canyon also is given, Hinds (34) has published an account of the melilite and nephelite basalts from Kauai and Niihau.

OLIVINE BASALTS

As shown in an earlier paper (37, pp. 298-310) the dominant Kauai lavas, both of the main complex and of the much later Koloa series, are olivine basalts, which apparently compose about 85 per cent of the total visible mass of the island. These rocks range in color from ash-gray, medium-gray, greenish-gray to dark-gray, geenish-black, brownish-black, or black. Texturally, most of the rocks are porphyritic, though many of the thin sills, narrow dikes, and a few flows are composed of aphanitic lavas. The most noteworthy feature of the olivine basalts is the size and abundance of the olivine phenocrysts which are visible to the eye or under a hand lens in at least 90 per cent of the rock bodies. Tabular feldspars are nearly or quite as abundant as the olivines, especially in the more salic varieties. Magnetite and augite, the other principal constituents, are visible in relatively few hand specimens. In structure, there are all possible gradations from highly scoriaceous or vesicular lavas to dense rocks, in which gas cavities, if present at all, are of microscopic size. While a few amygdaloidal flows are present, complete filling of the vesicles is rare. In many flows, the cavity walls are coated with zeolites, opal, quartz, or calcite. Calcite is being deposited in vesicles and fissures along the sea coast, and is found as linings or fillings of vesicles in flows and beach boulders along old shore lines. Flow structure is rare in the olivine basalts. The microscope shows that most of the rocks are holocrystalline except for occasional globules of brownish, isotropic, or partially devitrified glass. The abundance of glass naturally increases toward the top and base of the flows, though in most specimens examined, the texture is holocrystalline a very few inches below the surface. In a few thin sections, more or less continuous areas of glass or of its partially or wholly devitrified equivalent form an essential part of the groundmass.

In composition, the olivine basalts vary from types in which olivine is not conspicuous and plagioclase equals or exceeds augite in amount to pyroxenic facies in which olivine is abundant and plagioclase is distinctly subordinate. The end types are olivine andesite at the salic pole and ultra femic and picritic basalts at the femic pole.

Megascopically, most of the olivine basalts of Kauai can be separated into four groups which have proved of service for field classification:

1. Coarse grained chrysophyres: light to dark-gray lavas in which large, yellowish-green to deep, bottle-green olivines are common. Plagioclase tablets generally are abundant and many of them equal the olivines in size.

2. Fine grained chrysophyres: dense, dark colored lavas in which small, dark-green olivines are abundant. Feldspars are not so commonly phenocrystic as in the rocks of the first group. Both coarse and fine grained chrysophyres are found as flows, dikes, and sills.

3. Highly vesicular olivine basalts possessing saccharoidal texture: dark-colored, moderately coarse grained rocks with abundant phenocrysts of olivine and plagioclase. These are exclusively flow rocks.

4. Trappean basalts: dense, dark-colored lavas in which minute olivines may be visible under a hand lens. The thin sills, narrow dikes, and occasional flow rocks belong to this group.

The coarse and fine grained chrysophyres appear to compose 75 to 80 per cent of the olivine basalts of Kauai, and the proportion of the coarse to fine grained types is about 1 to 2. The highly vesicular, saccharoidal lavas are next in importance. The volume of the trappean rocks is small.

Other Types of Lavas

Andesitic basalts, containing less than 5 per cent of olivine and higher proportions of more alkalic labradorite than the normal olivine basalts, are next in abundance, but their total mass is relatively small. Pyroxene andesites (olivine-poor or olivine-free) are fairly common and are the most salic lavas so far found on Kauai. The andesitic basalts and the pyroxene andesites closely resemble the olivine basalts in megascopic appearance, except for the paucity of the olivine phenocrysts. A number of these more salic flows show well developed flow structure. Trachytic and other types of highly alkalic lavas, such as have been found on Hawaii, Maui, and Oahu, have not been recorded from Kauai. In addition to the types already mentioned, melilite basalt, nephelite-melilite basalt, nephelite basalt, analcite basalt, picrite basalt, picrite, diabase, and limburgite are represented in my collection. The nephelite, analcite, and the melilite basalts with one exception were obtained from the main igneous complex (34). A single melilitic flow, belonging to the late Koloa volcanic episode, was found west of Koloa on the south shore of Kauai. The melilite basalt described by Cross (7) is from the main complex and is not associated with a tuff cone of Koloa age as that writer believed. No hornblende-bearing or quartz-bearing rocks were found on the island.

Possible Plutonic Rocks

The presence of coarsely granular rocks on Kauai has been noted elsewhere (33, p. 22). Many of the flows contain gabbroid, pyroxenitic, and peridotitic xenoliths. Other coarse-grained rocks were found in pyroclastic deposits and as stream boulders whose sources were not ascertained. The list of petrographic types includes oligoclase diorite, oligoclase gabbro, olivine gabbro, olivine-free gabbro, troctolite, harzburgite, pyroxenite, and dunite. Whether these rocks were derived from deep-seated intrusions or from coarsely granular hypabyssal bodies cannot be determined in most specimens.

Recurrence of Petrographic Types

Of the various types of lavas, most were found to recur at various horizons in the Kauai dome, so that no sort of density stratification appears to exist. The nephelite basalts were found below elevations of 1,500 feet above sea level, but it is entirely probable that flows of this type are present at higher levels. Melilitic rocks were found from sea level nearly to the top of the dome. No difference appears to exist between the rocks of the older and younger Kauai series. The lavas of the late Koloa episode in general are more femic than those of the earlier volcanic periods, while highly plagioclasic types are distinctly less common. Highly differentiated facies, such as the melilite, nephelite, and analcite basalts, the limburgites, and the picrites were developed at various times during the principal volcanism, and are not confined to the late episode as Cross and Powers suggested.

Chemical Composition

Unfortunately, few analyses of lavas from Kauai are available; from Niihau several have been recently published by Washington and Keyes (59, p. 339). In Table 8, additional analyses are given.

TABLE 8. ANALYSES OF HAWAIIAN LAVAS

	1	2	3	4	5	6
SiO_2	48.99	45.48	47.24	40.56	48.69	49.73
Al_2O_3	13.73	11.87	12.80	8.81	14.00	13.71
Fe_2O_3	1.60	1.98	1.79	5.98	5.03	2.92
FeO	10.46	9.87	10.17	8.23	8.01	8.64
MgO	13.53	13.28	13.42	16.33	7.12	8.27
CaO	7.34	10.97	9.16	11.77	9.00	9.10
Na_2O	1.62	2.21	1.92	3.13	3.55	3.16
K_2O	0.27	0.77	0.52	1.19	1.24	1.02
H_2O+	0.27	0.74	0.50	0.27	......	......
H_2O-	0.10	0.23	0.17	0.06	......	......
TiO_2	1.73	1.90	1.82	2.73	2.29	2.84
P_2O_5	0.13	0.25	0.19	0.82	0.49	0.48
MnO	0.20	0.15	0.18	0.15	0.41	0.13
	100.38	99.94	100.24	100.03	99.83	100.00

1. Coarse-grained chrysophyre (Field group 1). Olokele Canyon, Kauai. W. T. Schaller analyst. Cross (7, p. 11).
2. Fine-grained chrysophyre (Field group 2). Olokele Canyon, Kauai. W. F. Hillebrand, analyst. Cross (7, p. 11).
3. Average of Nos. 1 and 2.
4. Picrite-basalt. Lihue, Kauai. H. S. Washington, analyst. Washington (58, vol. 12, p. 344).
5. Average of 43 analyses of lavas from the Hawaiian islands. Cross (7, p. 87).
6. Average of 56 analyses of the lavas of the island of Hawaii. Washington (58, vol. 6, p. 361)

It is impossible to estimate the average composition of the Kauai lavas from data now at hand, but because the olivine basalts of Field groups 1 and 2 are the dominant lavas of the island, I think that the average of the two analyses given in column 3, Table 8 represents a fair approximation of the olivine basalts. These rocks compose at least 85 per cent of the mass of the dome.

As Niihau is a part of the great, compound Kauai-Niihau volcano and its lavas are petrographically rather similar to those of Kauai, the analyses recently published by Washington and Keyes (59, p. 339) are presented. (See Table 9.)

Composition of Weathered Rocks and Soils

No weathered rocks from Kauai have been analyzed, but a few analyses of similar weathered rocks from Oahu have been published by Lyons (44). From a paper on the weathering of Hawaiian lavas (37) the following paragraphs have been taken.

As previously stated, the weathering of the Kauai lavas has taken place under a great variety of climatic conditions resulting from marked differences in rainfall over various portions of the island.

The weathering of the lavas has been accomplished primarily by chemical agents. Mechanical disintegration is important along the sea coasts, in stream channels, and on steep mountain slopes where landsliding and creep proceed at maximum rates, but the areas involved are limited. The effect of frost at higher elevations probably is of little consequence, since freezing temperatures develop only occasionally and the rocks in the highlands have been weathered chemically to very considerable depths below the surface. The chemical alteration of the lavas has taken place under climates ranging from rather arid to extremely humid; the mineralogical and chemical changes in the lavas vary with the climate of different sections of the island.

Unfortunately all of the analyses of Kauai soils and subsoils are of samples which have been taken from elevations of 1,000 feet or less, hence there are no representatives from the extensive, most heavily watered sections of the island.

The soils and subsoils of Kauai in the main are residual, having been derived from the decomposition in situ of flows and volcanic ejectamenta. Along the southern low-

TABLE 9. ANALYSES OF LAVAS OF NIIHAU

	1	2	3	4
SiO_2	49.73	46.44	46.86	46.36
Al_2O_3	14.56	16.21	14.78	14.77
Fe_2O_3	3.60	1.98	1.78	2.13
FeO	8.55	7.85	9.85	8.82
MgO	6.89	9.45	9.93	12.42
CaO	9.66	11.21	10.98	10.84
Na_2O	2.25	2.52	2.88	2.16
K_2O	0.62	.48	.40	.53
H_2O+	1.24	.54	.13	.57
H_2O-	.71	.10	.16	.12
TiO_2	2.21	2.11	1.67	1.31
ZrO_2	n.d.	......	n.d.	n.d.
P_2O_5	.22	.21	.20	.19
SO_3	n.d.	.11	n.d.	n.d.
Cr_2O_3	n.d.	.07	n.d.	n.d.
MnO	.11	.21	.15	.14
BaO	n.d.	n.d.	n.d.	n.d.
	100.35	99.49	99.79	100.36

1. Andesine basalt, Keanahuhi Valley, Kaeo Cone, Niihau. M. G. Keyes, analyst.
2. Olivine labradorite basalt. Kalaalaau Valley, Niihau. M. G. Keyes, analyst.
3. Olivine labradorite basalt, Puuwai Cone, Niihau. M. G. Keyes, analyst.
4. Olivine labradorite basalt. Nonopapa Landing, Niihau. M. G. Keyes, analyst.

lands, and to a less extent along the northern and eastern lowlands, fragmental volcanic deposits, erupted from local vents, have been superposed on the weathered lavas. These deposits, so far as can be determind from microscopic examination, are composed of fragments of the basaltic types previously described. Along the coast, detritus derived from the disintegration of local fringing reefs, is added to the basaltic débris. In the lower portions of the submerged river valleys on the lowlands of eastern Kauai, alluvial deposits, composed of basaltic detritus, have been formed, and represent the largest areas of transported soils. At the mouths of the rivers, reef limestone sand has been added to the basaltic alluvium. At the bases of steep mountain slopes are limited accumulations of material brought down by landsliding and creep.

The soils and sub-soils of the lowlands and drier mountain slopes of Kauai and the other Hawaiian islands are essentially residual laterites which vary widely in composition. . . . The brightest brick-red soils are to be found on the drier portions of the lowlands and the drier, low mountain slopes. With increasing precipitation, the soils become darker red, brownish-red, and brown. At varying depths below the surface, the pulverulent or clayey laterites pass rather abruptly into thoroughly decomposed but still compacted rocks which are brown, yellow, gray or green in color. The soils and sub-soils show none of the original textures or structures of the rocks from which they were derived; the weathered layers of lava below not infrequently have these more or less perfectly preserved. In the lower zones are great numbers of concentrically weathered boulders which often have at their centers spherical or ellipsoidal masses of relatively fresh rock from which the nature of the original lava can be determined. Still farther below the surface there is a gradual transition from the weathered to relatively unaltered rocks. The depth of the weathered zone apparently depends upon the amount of rainfall.

In the rainier sections of Kauai the dominant colors of the soils are brown, yellow, pink, and gray. These soils have lost a large part of their iron content and approach ferruginous bauxites in composition. The subsurface zones rapidly change in color to dark-brown or red. It is evident in the field that redeposition of iron oxide takes places abundantly below the bauxitic layers. Pulverulent masses of iron oxide fill practically every opening in the decomposed rock.

SOIL ANALYSES

The averages of soil analyses given in Table 10, compiled from data published by Kelley (41), doubtless present a very fair picture of the changes in composition which have taken place in the alterations of the basalts in the lowland areas of Kauai.

TABLE 10. AVERAGES OF SOIL ANALYSES

	1 AVERAGE OF 26 SOILS	2 AVERAGE OF 8 SUBSOILS	3 AVERAGE OF 26 SOILS AND 8 SUBSOILS	4 AVERAGE OF 10 RED SOILS	5 AVERAGE OF 13 BROWN SOILS
Insoluble residues	31.03	29.98	30.50	28.92	32.36
Al_2O_3	19.16	22.77	21.46	19.63	18.94
Fe_2O_3	21.50	20.57	21.03	15.27	18.75
MgO	1.32	.66	.99	.62	1.77
CaO	.60	.33	.46	.58	.56
Na_2O	.31	.36	.33	.34	.30
K_2O	.38	.31	.30	.32	.26
TiO_2	1.66	1.90	1.78	1.07	1.84
Mn_3O_4	.38	.42	.40	.30	.43
P_2O_5	.50	.40	.45	.57	.45
SO_3	.32	.32	.32	.36	.29
N	.12	.12	.18	.25	.24
H_2O	99.16	8.14	8.66	7.67	10.06
Volatile matter	10.05	14.57	12.31	14.97	15.06

A comparison of the average analyses of the 26 soils and of the 8 subsoils (Table 10) with the best average analysis of the Kauai lavas now obtainable shows the increase or decrease in the proportion of the various constituents to have been as shown in Table 11.

TABLE 11. COMPARISON OF AVERAGE ANALYSES OF COLUMNS 1 AND 2, TABLE 10.

SOIL		SUBSOIL	
INCREASE	DECREASE	INCREASE	DECREASE
Al_2O_3	SiO_2	Al_2O_3	SiO_2
Fe_2O_3	MgO	Fe_2O_3	MgO
FeO	CaO	FeO	CaO
Mn_3O_4	Na_2O	TiO_2	Na_2O
P_2O_5	K_2O	Mn_3O_4	K_2O
SO_3	TiO_2	P_2O_5	
H_2O		SO_3	
		H_2O	

Table 12, compiled from data published by Kelley (41), gives the mechanical composition of 4 soils (Nos. 1-4) from Koloa, southern Kauai, and of 2 soils and associated subsoils (Nos. 5, 6) from Hanalei, northern Kauai.

TABLE 12. COMPOSITION OF KAUAI SOILS

	VOLATILE MATTER	FINE GRAVEL	COARSE SAND	FINE SAND	SILT	FINE SILT	CLAY
1	18.75	.02	1.57	5.48	6.49	33.40	33.50
2	19.90	.10	3.11	7.58	7.22	36.50	27.40
3	15.60	1.23		11.78	11.65	24.20	35.80
4	13.74	.34		5.72	4.65	33.20	39.10
5	21.41	.10	.40	9.58	12.78	29.01	27.24
	16.42	.12	.12	11.88	9.04	22.28	41.09
6	25.28	.24	.42	18.35	10.31	32.45	15.04
	19.21	.12	.45	26.89	13.37	20.69	20.92

In Table 13 is listed the average of 8 analyses of alluvial soils from Hanalei Valley, as given by Kelley (41). The alluvial deposits in the lower portions of the submerged river valleys of the eastern half of Kauai are the principal accumulations of transported soils. They are composed of basaltic débris with considerable amounts of organic material. At the mouths of the rivers, calcareous sand is present in considerable quantities, and may be the dominant constituent of the soil.

TABLE 13. AVERAGE OF ANALYSES OF ALLUVIAL SOILS FROM KAUAI

Insoluble residue	44.57	TiO_2	2.62
Al_2O_3	18.81	Mn_3O_4	.27
Fe_2O_3	16.38	P_2O_5	.49
MgO	3.46	SO_3	.29
CaO	1.13	N	.19
Na_2O	.30	H_2O	....
K_2O	.17	Volatile matter	12.56

SEDIMENTARY DEPOSITS

The chief sedimentary deposits thinly veneer the coastal margins of Kauai, and have accumulated to depths of a hundred or more feet in the mouths of valleys crossing the emerged marine platform. Inland, there are local talus cones and aprons, stream gravels and boulders, and occasional small masses of iron oxides which have been precipitated in the pools of the highland swamps, but these inland deposits are of minor importance. The sediments for the most part are of Recent age; it is possible, however, that the basal sections of the discontinuous reef fringing the island and the lower horizons in the alluviated river mouths may date back to the closing stages of the Pleistocene, though no evidence on this point could be secured. The only sediments belonging to earlier stages of the island's history are: (1) boulder beaches which are found on the shoreward sections of the emerged marine platform; (2) beds of coarse conglomerate (the Waimea conglomerate) which are exposed near the base of Waimea and Olokele canyons at an elevation of about 500 feet. The ages of these older deposits could not be determined. Marine and terrestrial invertebrates, all Recent types, are found in the littoral deposits and in the sand dunes, and plant remains have been reported from borings made in the alluviated valley mouths. Unfortunately, none of the latter material was obtainable, so that specific identifications cannot be had. The results of microscopic studies of the various sediments have not been completed.

The sedimentary deposits of Kauai may be classified as follows:

1. Discontinuous, practically extinct, fringing reefs.
2. Clastic deposits of the littoral zone.
    a. Unconsolidated clays, sands, gravels, and boulders.
    b. Lithified beach sand and gravel (beach rock).
3. Emerged littoral deposits.
4. Sand dunes.
    a. Unconsolidated.
    b. Lithified.
5. Reef plain deposits.
6. Alluvial deposits.
7. Talus and landslide deposits.
8. Boulder beaches on the emerged marine platform.
9. Clastic deposits interbedded with the lavas of the main igneous complex.

The distribution of the principal areas covered by sediments is shown on the geological map of Kauai (Pl. XIII). As the various types of sediments merge into one another in many places accurate boundaries cannot be drawn.

THE FRINGING REEF

At various localities along the coast are patches of a narrow, practically extinct fringing reef (Pl. VII, *B*); these are naturally widest and most extensive on the windward (northern and northeastern) side, but even there are long stretches of reef-free coast. Off the steep northwestern cliffs only two small areas of reef are present, one near the mouth of Milolii Canyon and the other to the west of Nualolo Canyon; at a number of places, there are small colonies of reef-building organisms which can hardly be designated as reefs. An extinct fringing reef probably underlies the surficial deposits of the southwestern crescentic plain (Mana Flats).

The thickness of the Kauai reefs has not been determined. Some evidence may be gained from wells sunk at two places on the inner margin of the coastal lowland on the southwestern side of the island to obtain water for the Kekaha Plantation. Unfortunately, no log of the well was kept, and such samples as were taken have been lost. According to Mr. H. P. Faye, Manager of the Kekaha Plantation, the drills were driven "through the soil past boulders and into the solid rock" and continuous beds of lava were first encountered at depths of 250 to 325 feet below the surface (225 to 300 feet below sea level). The surface of the lavas is covered with a foot or two of soft material (probably weathered basalt), below which the drills went through 10 to 15 feet of very dense basalt. Below this dense rock, porous, fresh water-bearing lavas with occasional layers of dense material extended downward for 25 or 50 feet at which point the drilling was stopped. The water for irrigation is pumped from the porous horizons. No sedimentary deposits were encountered below the uppermost lava, which probably marks the surface of a submerged wave-cut platform upon which a reef grew or a limestone bank accumulated. This platform I correlate with the emerged marine plain on the eastern side of Kauai which was brought above sea level by the tilting of the island (p. 36). No information is available as to the nature of the boulders mentioned by Mr. Faye. Judging from the abundance of calcareous sand in the surficial deposits of the plain, an underlying reef limestone must have been its source. The boulders met in the drilling may be either limestone or masses of lava which have fallen from the cliffed highland behind the reef.

Between the mouth of the Wainiha River and Haena and at other places along the northern coast of Kauai, low plains have been developed by the deposition behind and over the reef of basaltic detritus from the highland and wind-blown, calcareous sand, derived from the disintegration of the practically extinct reef front and of shells lying along the shore (Pl. VI, *B*; IX, *A*). Along the shoreward margin, the surface soil is composed of calcareous sand, with small admixtures of olivine grains, basalt fragments, and weathered lava. Low dunes, rarely exceeding 30 feet in height, form a more or less continuous rampart along the shoreward edge of the plains behind the face of the reef (Pl. VII, *A*). Toward the base of the highland, the proportion of calcareous sand decreases, and basaltic detritus, including basalt and olivine sand, transported lateritic soil, coarse lava talus and stream lava boulders are the dominant components. The island is growing seaward at certain localities by this method of infilling behind the reef face. The surficial deposits of the southwestern (Mana) plain represent much more extensive accumulations which have completely buried the underlying limestone (p. 42).

The reefs about Kauai and the other windward Hawaiian islands are discontinuous and feebly developed. On Oahu, Setchell (54) found that the reefs are practically extinct. I observed the same condition on Kauai and Niihau. Apparently the reefs developed in late Pleistocene or Recent times in seas none too favorable for the reef builders; later they have been gradually killed off as certain elements in their environment became less satisfactory.

There is no evidence that pre-Glacial or inter-Glacial reefs existed about Kauai; the testimony of biologists is that the Hawaiian waters were not inhabited by corals until the close of the Pleistocene, but the principal components of the Hawaii reefs may not be corals. Certainly the present reef about Kauai is not growing on a subsiding basement. Kauai is not and apparently has not been a subsiding island.

The emerged, fringing reef on Niihau also has not grown up from a subsiding foundation. The northern, western, and southern portion of that island is a plain of low relief having a maximum elevation of about 125 feet at the base of the dome remnant which it borders. Most of this plain is surfaced by soil composed of basaltic and calcareous detritus. At a number of places, reef limestone is exposed at the surface; at others lava appears at or projects through the soil or limestone. Some of these lava masses are residuals of considerable size which were islands when the reef, underlying the surface soil, was growing about the island. Late extrusions of lava also have broken through the limestone in a number of places. At several localities, the contact between the limestone and the underlying lava basement is visible. The Niihau fringing reef apparently does not exceed 100 or 200 feet in thickness, and lies on a wave-cut platform which has a greatest width of about 8 miles. This island also remained relatively stable during the long period required for the erosion of the platform. Upon this the fringing reef flourished for a time and then the island was elevated.

The feeble character of the Kauai reef indicates its development in late Pleistocene or Recent times. If this assumption be correct, important sedimentary deposits would only be found along the outer margins of the platform, and the existing reefs would be merely thin pellicles rising from platforms which stand at no great distance below sea level.

For Kauai, the evidence for the existence of pre-Glacial reefs is negative. No remnants of pre-Glacial reef limestone are preserved on the emergent wave-cut platform, which is present around half the circumference of the island. Of course, the width of the platform has been materially reduced by wave attack since its emergence, so that it is possible that some reef material existed along its outermost borders. It would seem, however, that remnants should be preserved where perhaps the reef reached its maximum width or in some specially favored spot. It is evident from the extent of erosion of the platform that the emergence took place prior to the close of the Glacial Period. The erosion of the valleys crossing the platform below the present sea level is assigned to the periods of the lowered Glacial oceans; certainly it has not been accomplished entirely in Recent time.

The development of the tapering spurs between the gorges running into the formerly sea covered area at the base of Waialeale indicates the presence of a protecting barrier in front of them, but such a barrier is offered by Nonou and Kalepa ridges which parallel the eastern shore. Furthermore, some of the spur ends appear to have been slightly snubbed, thus indicating that wave attack was active within the protected area. The northern face of Haupu Ridge is strongly cliffed, and this may have been accomplished in part by the sweeping of the ocean waters through the wide gap between Haupu and Kalepa ridges. On the northern side of Kauai the marine platform abuts abruptly against the cliffed spurs of the northern face of the Anahola Mountains, showing that there had been no barrier to wave attack in this section. On the southern side, the erosional topography is largely obscured by the development of post-Glacial tuff cones so that definite evidence regarding the operation of wave action is difficult to obtain. The rise from the marine platform to the basaltic highlands is abrupt, however, and there is little indication of the development of tapering spurs between the valleys.

On the western side of the island, the marine bench is absent, and the highland terminates in strongly dissected cliffs of great height which either plunge into the ocean or are fronted by low, coastal plains. It is evident that these cliffs were developed during a period of long continued wave attack, which is correlated in part with that required to erode the wave-cut platform exposed on the eastern half of the island. The western cliffs continued to be eroded long after the elevation of the marine platform, but are now protected by the constructional plain. On the northwest, the great, coastal fault line scarp is still being actively eroded by waves. The dissection and retreat of the crest of the cliffs and the recent nipping of their bases indicate that wave attack at the present level was interrupted for a time—probably during the period of lowered Pleistocene oceans.

CLASTIC DEPOSITS OF THE LITTORAL ZONE

The clastic deposits of the littoral zone about Kauai may be divided into two principal groups: (a) Unconsolidated clays, sands, gravels, and boulders; (b) Beach rock (lithified beach sand and gravel).

Unconsolidated Clastic Deposits

The unconsolidated clastic deposits of the littoral zone vary from place to place, depending upon the nature of the coastal topography, the presence or absence of the fringing reef, and the proximity to river mouths. Along the cliffed, reef-free shores, the deposits are composed chiefly of coarse basaltic boulders, with subordinate quantities of finer basaltic and calcareous detritus, which in part is derived in situ and in part has been drifted from other localities by waves and currents. The boulders are derived from the cliff faces as the result of undercutting of the base by wave action and the slump of unsupported rock above; and by landsliding of material loosened by weathering. Rock masses also are torn from the lower portion of the cliffs by the powerful storm waves. The boulders vary greatly in size; in form they are predominantly angular or subangular. Where the river valleys enter the ocean along the cliffed coast, fluvial deposits, which also may contain great boulders, are added to those of marine origin. The river-borne boulders are more or less rounded; the finer sediment is composed of basaltic sand and chemically weathered basaltic material. At many of the river mouths along the cliffed coasts, deposits of finer texture, consisting primarily of calcareous sand with smaller amount of basaltic débris, are frequently formed. The calcareous sand has been drifted along the cliffed coast from the limestone deposits of the windward portion of the island, and has lodged in the various indentations along the shore line. Few of these sand beaches last more than a few years, as storm waves sweep them away from time to time. New deposits may be formed within a short period, or may not be developed for several years.

The finer-textured deposits of the littoral zone are dominantly arenaceous; argillaceous sediments are relatively unimportant, except at the mouths of streams which are transporting only fine detritus. Even at such places, considerable quantities of calcareous sand are usually mixed with the clays. The littoral psammites are dominantly calcareous, but their composition varies within wide limits due to the presence of more or less basaltic debris. Four principal types may be recognized: 1, dominantly calcareous sands; 2, dominantly basaltic sands; 3, intermediate sands; 4, olivine sands.

The olivine sand beaches are less abundant than the others. Olivine is the dominant constituent, but basaltic and calcareous grains are also present in varying amounts.

Most of the fine grained littoral sediments are temporary deposits lasting for comparatively brief periods. The severe storms of the rainy seasons not infrequently wipe out the accumulations of several years, or greatly alter the form and size of the deposit. Such changes are particularly noticeable at the reef-free river mouths and along the cliffed coasts.

Beach Rock

At a number of localities along the shores of Kauai the calcareous sands have been cemented into a more or less firmly indurated limestone to which the term "beach rock" is applied. The limits within which lithification takes place apparently are the high spring tide level and a depth not more than 10 feet below low tide level. Cementation of the loose calcareous sand takes place most readily along open coasts, where heavy surf beats more or less continuously against the shores. To a less extent, the process is in operation in partially protected localities, such as the shores of Nawiliwili Harbor, and behind the reef at Haena. Each of these localities, however, is subjected to powerful wave action, especially during the rainy season. The beach rock is composed primarily of calcareous sand, though small amounts of fine to medium-coarse basaltic débris are always present. In places, large basaltic boulders are included in the beach rock.

The sands about the shores of Kauai have been swept up from shallow depths off shore by storm waves. Powerful wave action apparently is necessary in the process of cementation, which takes place only a short distance above and below low tide level. The shallow depths of water at which beach rock was observed are constantly affected by movements of the surf, and very violently so during heavy storms. There is no evidence that the beach rock observed below low tide level was formed above sea level and later submerged; in fact, the most recent movements of the island have been in the opposite direction.

Pitted Shore Lavas

Where vesicular lavas are present along the shore line, the surfaces of wave-cut benches and of beach boulders are usually strongly pitted. The cavities are circular or elliptical in ground plan, and have nearly vertical walls. In size, they vary from small depressions little larger than the vesicles from which they apparently developed to tubes 6 to 7 inches in length and 2 to 3 inches in diameter. The pits are formed in part by solutions which have attacked the vesicle walls and in part by mechanical abrasion by sand and gravel lodged in the cavities and kept in motion by waves. Many lava flows along elevated shore lines and boulders of the emerged beaches show this pitting. Along the shore, marine animals, especially the echinoid (*Podofera pedifera*) and the limpet (*Helcioniscus exaratus*) make their homes in the cavities. Pitted boulders and surfaces are occasionally found in the lavas at various elevations throughout the island, but the cavities differ from those of the shore lavas in that they are more irregular in ground plan and generally are much shallower. Such depressions appear to have been developed by the enlargement of vesicles resulting from the action of ground water and atmospheric solutions. Mechanical abrasion apparently has not been important.

EMERGED LITTORAL DEPOSITS

At a few localities along the eastern shore of Kauai, littoral sediments have been exposed as the result of the Recent slight emergence of the island. The principal deposit forms a more or less continuous belt that extends from just north of Hanamaulu Harbor to a point about three-quarters of a mile south of Kealia; a second deposit lies between Kealia and Kapaa; and a third deposit of different character has been exposed by the building of the government road on the south side of Nawiliwili Harbor.

The deposits between Hanamaulu Bay and Kealia are composed of basalt and reef limestone boulders, finer basaltic detritus, calcareous sand, and abundant remains of marine invertebrates, embedded in a brick-red, sticky, lateritic clay. The deposit forms discontinuous lenses which vary in thickness and in the proportion of their constituents within relatively short distances. The exposed thickness of the series nowhere exceeds 6 feet. These sediments lie either on a somewhat weathered basaltic surface or upon nonfossiliferous, lateritic soils. At no place is the top of the series more than 15 feet above high tide level. The deposits are similar to those which are being formed at the present time at many places along the shore line.

The following list includes the fossils collected from the deposits between Kealia and Kapaa (column 1) and north of Hanamaulu Bay (column 2). The species have been identified by Professor Junius Henderson; all of them are Recent.

	1	2		1	2
Coral fragments, Fungia Porites	X	X	Bittium, several undetermined sp.	X	
Undetermined echinoid spines	X	X	Eulima major Sowerby		X
Anemia nobilis Reeve	X		Helcioniscus argentatus Sowerby	X	X
Chama sp.	X		Helcioniscus exaratus Nuttall		X
Spondylus sp.	X		Hipponyx antiquatus Linneaus	X	X
Cadakia ramulosa Gould		X	Melampus sp. undetermined	X	
Venus reticulata Linneaus	X	X	Nerita picea Recluz		X
Conus hebraeus Linneaus	X		Patella sp.		X
Conus lividus Hwass	X		Purpura aperta Blainville	X	
Conus striatus Linneaus	X		Ricinula ricinus Linneaus	X	
Conus sp.	X	X	(Possibly Drupa sp.?)		
Cypraea caput-serpentis Linneaus	X	X	Scalaria sp.	X	
Cypraea isabella Linnaeus	X	X	Strombus maculatus Nuttall	X	X
Cypraea mauritiana Linnaeus	X		Triton cf. rubecula Linneaus	X	
Cypraea madagascariensis Smith		X	Trochus sandwicensis Souleyet	X	X
Cypraea reticulata Martyn	X	X	Turbo chrysostomus Linnaeus	X	
Cypraea sp.	X	X	Undetermined gastropods	X	X

A third emerged beach deposit extends for perhaps a third of a mile east of the mouth of the Huleia River, along the Government breakwater road on the south side of Nawiliwili Harbor. This deposit is composed of dark, brownish or brownish-black, medium to fine grained sandy soil in which a considerable number of marine invertebrate shells are embedded. The top of the series apparently is about 25 feet above sea level. The shells collected from this locality unfortunately were lost. According to my notes, they include a number of species similar to those now living about the shores of Nawiliwili Harbor. The species appeared to differ somewhat from those listed above, and properly should, because the sands were accumulated in an estuary and not along the open ocean coast.

SAND DUNES

Unconsolidated Sand Dunes

Low sand dunes, few of them exceeding 30 feet in height, are present along most of the coastal margins of the southwestern lowland (Mana Flats) and the hinter-reef plains of other parts of Kauai, where there is a rapid accumulation of disintegrated limestone fragments by wind and wave action (Pl. VII, *A*). They are composed primarily of calcareous sand, but small admixtures of basaltic débris are present. Slight lithification was observed in certain of the deposits. Some fossiliferous horizons are present. In the dunes along the coast west of Haena there are some thin layers of blackish, basaltic soil mixed with more or less calcareous sand, which contain abundant shells of land snails. Otherwise, the fossils are of marine invertebrates.

The following species from the dunes west of Haena were identified by Professor Henderson:

MARINE INVERTEBRATES

Echinoid spines (Heterocentrotus sp.)
Chama sp.
Cadakia ramulosa Gould
Mytilus crebistriatus Conrad
Perna californica Conrad
Perna costellata Conrad
Venus reticulata Linneaus
Cerithium obeliscus Bruguière
Conus lividus Hwass
Conus hebraeus Linneaus
Conus sp.
Cypraea caput-serpentis Linneaus
Cypraea reticulata Martyn
Helcioniscus argentatus Sowerby
Helcioniscus exaratus Nuttall
Modulus tectum Gmelin
Nerita picea Recluz
Neritina granosa Sowerby
Patella sp.
Purpura aperta Blainville
Ricinula ricinus Linneaus
Sistrum morus Lamarck
Strombus maculatus Nuttall
Triton pilearis Linnaeus
Triton tuberosa Linnaeus
Trochus sandwicensis Souleyet
Turbo chrysostomus Linnaeus
Vermetus sp.
Small,undertermined gastropods.

LAND SNAILS

Carelia dolei Ancey
Lyropupa scabra Pilsbry and Cooke
Lyropupa rhabdata baldwiniana Cooke and Pilsbry
Nesopupa dispersa Cooke and Pilsbry

Lithified Dunes

Along the southeastern shore of Kauai, east of Koloa, certain of the older calcareous sand dunes have been lithified, and later have been partially eroded by wave action, with the result that cliffs 20 to 25 feet in height have been cut into their seaward margins (Pls. IX, *B*, *C*). The dune sand is thinly laminated, and the cross-bedded structure is prominently developed. The stage of induration varies from place to place. The surficial layers have commonly been so thoroughly lithified that the granular texture has been practically obliterated. Solution at the surface has caused deep pitting; the intervening areas have commonly been reduced to sharp pinnacles of extraordinarily resistant limestone. Dana (19) who first described these lithified dunes, believed that the compact surface layers had been produced by the deposition of limestone in the depressions in which water commonly stands. Microscopic examination shows, however, that the granular structure is visible in the lithified crust; therefore, this crust has been formed by the deposition of calcium carbonate about the grains of the sandstone not only in the depressions, but also on the intervening areas. The induration of the material composing the pinnacles would hardly be as great as that of the floors of the depressions had the process operated as outlined by Dana. It appears that the cementation of the first inch or so of the surface layers has taken place by the solution of calcium carbonate at the surface, and its rapid deposition almost immediately below the surface.

SURFICIAL DEPOSITS OF SOUTHWESTERN LOWLAND AND HINTER-REEF PLAINS

In the deposits surfacing the southwestern lowland (Mana Flats) and the hinter-reef plains, a shoreward margin of calcareous sand is everywhere present. (See p. 42.) The sand is composed of fragments of the reef limestone and of shell fragments. The width of the calcareous margin varies, but is generally at least a third of the total width of the plain. Inland the percentage of basaltic detritus increases; on the wider plains, such as the southwestern lowlands, there is very little calcareous material at the base of the cliffs, but on the narrower plains, the amount of calcareous sand along the inner margins is relatively large. The basaltic débris consists largely of lateritized material, which has been transported from the highland by the streams which debouche upon the plains. Grains of unaltered basalt and of rather fresh olivine and magnetite are also common. The basaltic soil cover is generally brownish, dull red, or black in color. When mixed with considerable amount of calcareous material, the soils have a speckled appearance. At the base of the cliffs, an apron of talus and alluvial deposits usually is present.

On the northern side of Kauai, the extinct fringing reef projects out for some distance beyond the detrital deposits; on the southwestern side, the materials have completely buried the underlying limestone.

ALLUVIAL DEPOSITS

The principal alluvial deposits occupy the lower portions of the stream valleys which cross the emerged marine platform (Pl. V, *A*). Deposits of less extent are present at various places in all of the stream channels and at the mouths of the valleys opening from the highland upon the plains.

The sediments of the submerged valley mouths consist primarily of reddish, brownish, and blackish lateritized soils with which are mixed minor quantities of more or less unaltered basaltic débris and a considerable amount of organic material. Upstream, these deposits merge into coarser sands and gravels, which floor portions of the valleys to slight depths. Wells bored in the bottom of Hanapepe Canyon about two miles from the shore passed through more than 100 feet of alluvial material, the lower 75 feet of which is below sea level. Borings for the supports of the bridge which crosses the mouth of the Wailua River were sunk through at least 80 feet of alluvium without reaching the underlying lavas. No exact records were kept of any of these borings, but according to Mr. J. Moro, who drilled the Hanapepe wells, certain of the beds passed through are as follows: "Dirt, soil, and river gravels—about 40 feet; logs were met in the gravels about 30 feet below the surface. At a depth of 50 feet is a hard, red soil, below which is porous soil. Lavas were first met at depths of 100 to 108 feet below the surface." In the mouth of the Wailua River, the deposits are described as sand, gravels, and clays. Masses of plant remains were encountered at some depth below the surface.

At the mouths of the streams, large quantities of calcareous sand are mixed with the laterized soil, so that the sediments become dominantly calcareous. A calcareous sand bar or spit is usually present at the mouths of the valleys, and across this the river maintains a relatively narrow connection with the ocean.

The formation of the alluvial flats has followed a submergence of the valley mouths. It has been suggested (pp. 38-39) that the streams eroded their channels below the present sea level when the oceans were lowered by the formation of the Pleistocene ice sheets; the submergence and consequent alluviation took place as the oceans returned to their normal levels. If this explanation be correct, the basal deposits in the valley mouths date back to the Pleistocene. As sedimentation continued the upstream portion gradually rose above sea level; the Recent slight emergence of the island has caused a further down stream extension of the subaerial phases, so that extensive areas are now flood plain and channel sediments veneering marine accumulations below. Branner (4) has described the alluviated mouths of certain great canyons on the eastern side of Kohala, Hawaii, and has attributed the submergence to a sinking of the mountain.

Above the alluviated portions of the marine plain valleys, the fluvial deposits are chiefly gravels and boulders. In the highland canyons, the deposits are much coarser in texture, and toward the heads of the canyons, are wholly made of very coarse gravels and boulders.

TALUS AND LANDSLIDE DEPOSITS

Because of the extreme ruggedness of the relief of Kauai, talus deposits are abundantly present. The great talus-alluvial aprons at the base of the highland cliffs along the hinter-reef plains of southwestern, western, and northern Kauai are the most notable examples. In places they exceed 200 feet in height. Talus cones and small landslide deposits are common in all of the deep canyons.

BOULDER BEACHES ON THE EMERGED MARINE PLATFORM

At a number of localities, particularly on the eastern section of the emerged marine platform, boulder beaches of considerable longitudinal extent are exposed. The deposits are composed of subangular or angular boulders of varying sizes; few of then exceed 3 feet in diameter. Boulders of vesicular lava commonly show pitting similar to that being developed in the vesicular lava along the present shore line. (See p. 70). No recognizable sediments of any type are associated with the boulder beaches. Apparently as the sea retreated during the emergence of the island, the finer, unconsolidated detritus was gradually swept seaward, and only the larger rock masses were left. The distribution of the beaches is rather difficult to ascertain, owing to the fields of sugar cane which obscure their relations, and to the fact that the boulders have been removed from many places in order to render the land available for cultivation. It seems, however, from a study of undisturbed deposits along the shoreward section of the eastern lowland, that there are a series of beaches, each marking a temporary still stand of the island during the period of emergence. The boulders in certain places are definitely related to minor wave cut terraces which were eroded into the platform during the course of the emergence.

The evidence favoring the accumulation of the boulder deposits by wave action is; (1) their distribution over the wave-cut platform; (2) the arrangement of the deposits where undisturbed, parallel to the present shore line; (3) the character of the pitting of the vesicular boulders. The age of the beaches is unknown, because the period of the emergence of the island cannot be definitely dated. The deposits antedate the basal alluvial sediments in the mouths of the valleys crossing the plains and the cutting of the valleys themselves. As it is considered that the valley alluvium in part is Pleistocene, the boulder beaches are older Pleistocene or even pre-Pleistocene in age.

CLASTIC DEPOSITS INTERBEDDED WITH LAVAS

In the bottom of Waimea Canyon near the mouth of the Poomau, and in Waialae and Koaie canyons, which also join Waimea Canyon, a coarse conglomerate is exposed. The deposits are composed of moderately well rounded boulders up to 3 feet in diameter mixed with gravel and sand. The whole is rather well indurated. The deposits are of sedimentary origin, and represent local accumulations upon an erosion surface during a rather long interruption in the volcanic cycle. Though the lower contact is exposed in only a few places, I think the evidence is sufficient to warrant the separation of the Kauai lavas into two groups at this break. (See pp. 56-57). I was not able to trace the conglomerate into other canyons, but this may be due to original local deposition, depth of rock weathering or to the heavy cover of vegetation.

DIASTROPHIC MOVEMENTS

CHANGES OF LEVEL

The tilted emerged platform, the submerged platform, and the submerged valley mouths provide abundant evidence that the Kauai dome has been subjected to notable changes of level at different times since the close of the last principal volcanic period (the late Kauai volcanicity).

The broad platform of the eastern half of Kauai (Pl. VI,*C*) is a wave-cut bench which was elevated by a tilting of the island along a northeast-southwest axis running from Hanalei Bay to the mouth of the Waimea River, (See pp. 35-36). That the emergence was not continuous is shown by the presence of a number of minor terraces which were cut into the platform during short periods of still stand. These subordinate benches are best developed on eastern Kauai where at least three can be recognized—one at 30 feet above sea level, a second at 50 feet, and a third at 100 feet. Other terraces and beaches may be present, but the sugar-cane fields which cover much of the platform obscure minor topographic features.

The principal elevation of Kauai does not seem to have been contemporaneous with that of Niihau. A wide, wave-cut platform borders the dome remnant of Niihau on the north, west, and south, but it does not stand so high above sea level as does the Kauai platform nor is there evidence of tilting during the course of its elevation. Much of the Niihau platform is covered by a thin fringing reef. As preglacial or interglacial reefs do not seem to have existed in the Hawaiian Archipelago, I consider that the elevation of Niihau took place late in the Pleistocene or in early Recent times after reef building organisms had found a foothold about the islands. Niihau therefore appears to have been elevated much later than Kauai. As these two mountains are parts of a single great volcanic structure, their different behavior requires explanation. The profound faulting which caused the disappearance of the eastern half of Niihau virtually broke the compound volcano in two, and subsequently the two sections behaved as more or less independent units. During the course of the faulting, the eastern or Kauai block was tilted toward the northeast and the emergence of the platform took place; the western side of the block was further submerged below sea level. The elevation of the Niihau block occured considerably later.

SUBMERGED WAVE-CUT PLATFORM

The evidence of the existence of a submerged platform along the western side of Kauai is meager enough. (See pp. 35-36). Records of wells drilled near Mana and Kaunalewa show that the drills passed through surficial soils and deposits of boulders for 275 to 300 feet below sea level, and then through a thin soil layer and into solid lava. Apparently the boulders overlie a wave-cut platform slightly weathered at the surface. This platform I consider to be continuous with that of the eastern two-thirds of the island. Probably it is not present at the above depths along the northwestern (Na Pali) coast, as faulting seems to have occurred there and a platform, if present, would have been carried to much greater depths below the sea level.

The mouths of the valleys crossing the platform and of the larger highland canyons have been cut at least 100 feet below the present sea level and subsequently have been submerged and partially alluviated.

RECENT CHANGES OF LEVEL

The proofs of recent changes of level on Kauai are: (1) wave-cut terraces—narrow rock benches at the base of the recent cliffs which have been cut into the emerged platform or into the base of the recently nipped headlands; (2) emerged sea caves; (3) emerged littoral deposits of various types.

Narrow wave-cut terraces 4 to 20 feet above mean sea level are present about the shores of Kauai, except off the constructional plain on the southwestern side. The main bench stands about 4 to 6 feet above sea level, though it varies somewhat in elevation from place to place (Pls. VI, *A*, *C*; VIII, *B*). Its average width is 20 to 30 feet, but in places exceeds 150 feet. The variations in elevations are caused by lithologic variations in the lavas, the presence of adjacent vesicular and nonvesicular areas, jointing, and variations in the dip of the flows along the sea coast. The most even-surfaced bench has been cut in the tuff cone and near-by sea stacks composed of the same rock north of Kilauea, northern Kauai. The terrace is definitely wave-cut; it transgresses the dip of the lavas and tuff beds into which it has been eroded. Other narrow benches are present up to 20 feet above sea level, but they are not continuous for any distance about the island shores. Most of these are structural benches; they have about the same dip as do the flows. Some of the higher benches appear to truncate the flows, and may be of wave cut origin, but the evidence is not conclusive .

Further evidence of this recent change of level is providel by emerged sea caves which are fairly common along the steeply cliffed coasts of Kauai. Unfortunately, most of these caves are inaccessible because of the difficulty of landing. One to the south of Carter Point on the southeast side of the island was studied. This cave is 75 feet long, 79 feet wide at its mouth, and 35 feet wide at the back. The height of the mouth is 22 feet and of the back wall 8 feet. The lip of the cave now stands about 10 feet above mean sea level, the base of the back wall slightly over 20 feet. The lips of some of these emerged caves appear to be a little higher above sea level but accurate determinations could not be made. In the caves which are now being excavated, the lips appear to be 6 to 10 feet below mean sea level, hence the amount of emergence indicated by the dry caves is from 16 to 20 feet.

Recently elevated beach deposits are present at various places between Hanamaulu Bay and Kealia. The tops of these beaches are not more than 15 or 18 feet above sea level. The emergence of these deposits certainly is to be correlated with the recent change of level, because the fossils found in them are of types living about the shores of the island today. The sediments were formed just above and below sea level, consequently they now stand at higher elevations than the terraces which were cut at depths of 8 to 12 feet. The change of level indicated by these deposits is about 15 feet.

Wentworth and Palmer (63) have described terraces standing 4 to 12 feet above sea level on all of the Hawaiian islands and on other islands in the North Pacific. In conjunction with similar evidence obtained in other parts of the world, these writers believe that a negative shift of ocean level of 12 or 15 feet has taken place. In his recent paper on the geology of St. Helena Island, Daly (17, pp. 80, 81) gives the figure as 15 or 16 feet.

FAULTING

It is apparent that extensive faulting has taken place on Kauai, but exact location of the faults and the amount of displacement along them is practically impossible to determine because of the difficulty of tracing the dislocations in homogeneous rocks, the great amount of weathering and of erosion which have occured, and the heavy cover of vegetation in many critical places. The principal movements along the faults were completed early in the erosional period; no evidence of recent movements was observed. The most certain faults are indicated in figure 12; others undoubtedly are present, but the evidence for their existence is insufficient to justify mapping. The evidence for faulting is almost entirely physiographic; actual traces of fault planes were found in only a few places, and these unfortunately are on insignificant fractures.

The clearest example of extensive engulfment is the great volcanic sink between Anahola Mountains and Haupu Ridge, the development of which does not seem possible on any other hypothesis. (See p. 29). The depression may represent the enlargement of a summit sink of the Kilauea type which existed during the volcanic period, or it may have originated after

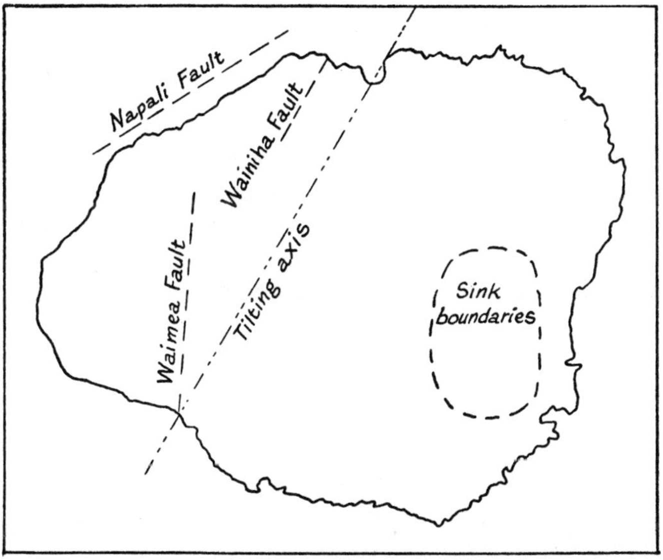

FIGURE 13.—Fault map of the island of Kauai. The positions of the faults were determined principally from physiographic evidence and consequently are only approximate. The axis along which the tilting of the island took place is also shown. Scale: 1 inch=9.7 miles.

all eruptivity had ceased; no evidence favoring either view was obtained. It is impossible to locate the faults bounding the depression, but I have the impression that the relations were similar to those of Kilauea, though on a larger scale. The present dimensions of the floor are 9 miles in a north-south direction and 6½ miles in an east-west direction. The size and depth of the sink may have been increased by fluvial and marine erosion which went on after the engulfment. The faults indicated on the map (fig. 13) as marking the present margin of the floor may not be accurately located. They are drawn merely to indicate that the depression is of tectonic origin.

Dana (19) and also Powers (52) have described the great line of cliffs on the northwestern side of Kauai (the Na Pali coast) as a battered fault scarp, and certain evidence, such as the straightness of the cliff line and important breaks in the submarine contours favor this explanation. The present cliffs constitute a fault line scarp rather than a fault scarp; wave action certainly has caused some retreat from the line or zone of dislocation, but the amount of this retreat can not be determined from a few soundings which have been taken off this coast. To the east and to the west of the fault line scarp, the cliffs are of marine origin.

The aberrant course of the Waimea River seems to have been determined by a fault zone traversing the island in a north-south direction. (See p. 36). The block on the west stood higher than that on the east, and consequently deflected the drainage from the summit plateau, and made it tributary to a north-south trunk stream.

Faulting also seems to have controlled the location of parts of the courses of other streams on Kauai (for example, the lower portion of the Wainiha), but the evidence is so uncertain that no proof can be advanced in support of this statement. Other fracture lines are indicated by the linear distribution of some of the cinder cones, but they have produced no other topographic expression.

The amounts of displacement along these fault zones cannot be ascertained.

RELATIVE AGES OF HAWAIIAN LANDSCAPES

IS KAUAI THE "OLDEST" HAWAIIAN DOME?

In literature dealing with the natural history of Hawaii, Kauai is commonly described as the "oldest" island of the windward section; that is, the close of the last principal eruptive period there first took place. Because this island has suffered the most extensive modifications of its initial outlines, the argument is advanced that the posteruptive history must have been of longer duration than on any other dome in the windward group. Also, evidence derived from the highly specialized nature of the flora and fauna and from the depth of rock weathering has been cited in favor of this view. Dana (19; 21, p. 311), states that "the valleys of Kauai are as much more extensive than those of the other islands of the group as its peaks are more irregular, abrupt, and broken" and that "the valleys and peaks [of Kauai] indicate that its fires long ago ceased,—as long ago as those of eastern Oahu, if not before." In order of extinction of the principal Hawaiian vents, Dana places Kauai first.

Dutton (23, pp. 83, 84) says of the older domes in Hawaii

> . . . we have no means of judging the antiquity of their final action except by the progress made by erosion in demolishing them, and this progress is, in every instance, considerable. It is most conspicuous on Kauai and Oahu, and almost equally so on Molokai and west Maui. From this it is inferred that the western islands of the group have longest enjoyed immunity from eruption. Kauai, especially, is frequently spoken of as the oldest of the group, and judging from the amount of destruction wrought upon it by the eroding forces, the statement is in some measure justified—but only to this extent: the period which has elapsed since the cessation of eruption has probably been longer than in any of the other islands. It does not follow, however, that the eruptions began here any earlier than on Hawaii. Whether they did or not is a question which I see no way of determining.

Among later writers Hitchcock (40, p. 14), Cross (7, p. 9), Powers (52, p. 501), Bryan (5), Martin and Pierce (45, p. 37) and Wentworth (62, pp. 131, 132) have considered that the extreme dissection of Kauai establishes its greater age.

None of these investigators seems to have studied the morphologic development of the Hawaiian domes in general or that of Kauai in detail; else different conclusions must have been reached. Also, none has considered the varying rates of marine and fluvial erosion in different parts of the islands, a vital factor in estimating the ages of the present landscapes. Furthermore, the importance of faulting in causing modifications of the initial outlines has largely been lost sight of.

EVIDENCE APPARENTLY FAVORING GREATER AGE OF KAUAI

Certain lines of evidence seem to favor an earlier close of the last principal eruptive period on Kauai:

1. The topographic map shows that Kauai has suffered somewhat greater erosional modifications than has Oahu, West Maui, Lanai, Molokai, and the Kohala section of Hawaii.

2. The principal streams of Kauai are more numerous and are more permanent and less variable in their flow than on any of the other mountains; a longer time is inferred for their development. With the exception of a few streams on windward (eastern) Mauna Kea, Hawaii, those of Kauai have the largest average volumes.

3. Kauai has suffered extensive inroads from wave attack, but the evidence as yet is insufficient to prove whether greater destruction has been wrought than on Oahu or Niihau. The origin of the cliffed shores of these islands and the lengths of the time which have been required for their development need further investigation. The cliffed coasts of northeastern Molokai, eastern Niihau, and the northwestern coast of Kauai are considered fault scarps or fault line scarps. The cliffs of western Kauai, western Niihau, northern West Maui, southeastern Molokai, western Molokai, and eastern Mauna Kea, Hawaii, certainly have been cut by waves.

4. Botanical and zoological evidence seems to favor the greater antiquity of the present Kauai landscape. Hillebrand (30, pp. XVIII-XXII), the pioneer in botanical work in Hawaii, writes:

> As the age of the various islands increases in progression from east to west, it may be inferred that the richness in endemic species will stand in the same ratio. Kauai is not only richest in species but has them on the whole more differentiated.

Bryan (5, pp. 103, 193), apparently quoting from Hillebrand, states:

> . . . the number of endemic plants . . . is largest on Kauai and the smallest on the large island of Hawaii. . . . Disintegration of the lava has proceeded farther here (on Kauai) than on the other islands, a fact taken in connection with other data as indicating that the volcanic fires died out first at this end of the chain.

From observations on other Hawaiian islands, I consider the validity of the statement regarding the deeper weathering of the Kauai lavas doubtful; at least before it can be established much more evidence must be obtained from all parts of the group. In Hawaii, the depth to which rock alteration has proceeded is not necessarily a measure of the age of the lavas, but rather of the amount of rainfall in a given area and the ease with which the groundwater solutions can penetrate the rocks. On the various domes, the zone of weathering is deepest on the rainy windward flanks and in these sections varies with the rainfall.

According to Hitchcock (40),

the confirmation of our belief in the greater antiquity of Kauai over the other islands is derived from the study of plants. . . . Taking the extremes, it may be stated that the flora of Mauna Loa is the poorest and most uniform and that of Kauai is the richest and most individualized in species.

From studies of the distribution of the Achatinellidae in Hawaii, Pilsbry (49) has concluded that all the Hawaiian islands once were united into a single land mass: "Volcanic activity built up the older masses, subsidence followed, Kauai being the first island dismembered from the pan-Hawaiian area."

Campbell (6, p. 268) writes: "The flora of Kauai is not only richest in number of species, but the species are most highly specialized and many are found exclusively on this island, indicating an early separation of the island from the other members of the archipelago."

There seems to be no geological evidence favoring such a former continent or series of continental islands. Biological necessity alone has called it into being.

In regard to the biological evidence, the question may be raised whether the greater abundance and differentiation of the Kauai species has taken place because of the longer time available or because of an environment more favorable then on the other mountains. The least diversity of the flora and fauna naturally appears where volcanic activity has longest prevailed. This subject has not yet received sufficient investigation to accept the biologic evidence as final in the problem of the age of the islands.

5. The fact that the migration of volcanic activity in the Hawaiian range has been in general from west to east is significant, but it does not establish the exact order of cessation of the last principal volcanic epoch at the various centers. The Kohala dome of Hawaii, for example, is associated with two active and two recently extinct domes; yet Kohala is older than eastern Maui and possibly also other domes lying to the west. The sealing of the principal eruptive areas therefore has not taken place in orderly fashion from west to east.

On the doublet islands, Oahu, Molokai, and Maui, the western member of each became extinct before the eastern, and has been partially buried by the lavas of the younger volcano. If this relationship holds for the Kauai-Niihau doublet, Niihau which lies to the southwest of Kauai, is the older of the pair.

The evidence which has been cited seems to favor the greater age of the present Kauai surface, but, in all of the morphologic discussions, the highly important factor of the rates of weathering and erosion on the various domes and on different sections of the same dome has received practically no consideration. Furthermore, insufficient attention has been paid to the loss of mass of the cones resulting from engulfment.

RELATIVE AGE OF KAUAI

In determining the length of time required for the erosional modification of originally similar land forms which are composed of rocks of about the same resistance to weathering and erosion and varying only slightly in their structural details, the degree of dissection of areas which may be selected for comparison will represent a fairly exact measure of the time intervals provided the rates of degradation prevailing over the various areas are approximately equal. On the other hand, because physiographic stage is not necessarily a measure of age, the dissection of regions in the same stage of the erosion cycle may have required vastly different lengths of time, hence the factors controlling the rate of erosion must be examined in detail. In areas such as the Hawaiian Archipelago, where the rocks are more or less uniform in their resistance to weathering and erosion, and where the structures of the mountains show few variations, the speed of weathering and erosion becomes the determining factor in the amount of removal which takes place within a given period of time. There are notable differences in rainfall over the various domes and over different sections of the same dome which affect the number, size, and fluctuations in volume of the streams. (See p. 46). There the rates of erosion over different sections of the same mountain and over similarly located sections on the various mountains are far from uniform. Domes which became extinct at approximately the same time have not been eroded to the same extent, and certain domes which have been long extinct have been less deeply dissected than those on which the last major volcanic period prevailed considerably longer.

The windward sections of all of the volcanoes have been more deeply eroded than the leeward; the constructional surfaces of the older domes have been almost completely destroyed on the windward side and are much broken on the leeward. On the younger mountains, conic sections of the constructional surfaces separate the windward canyons, while the leeward slopes have been little dissected. The summits of the highest peaks, Haleakala, Mauna Loa, and Mauna Kea, project above the zone of maximum rainfall, hence they are almost unbroken by erosion. On the low islands, Niihau, Lanai, Kahoolawe, and western Molokai, because of the light rainfall there are no permanent streams, and the rate of fluvial erosion has been extremely slow; changes wrought by wave action are much more important.

It is evident that engulfment also has greatly altered the form of certain of the domes; this subject has received insufficient consideration in papers dealing with the morphology of the islands. The loss of bulk on the part of a lava dome is not necessarily a measure of the amount of erosion which has taken place, as, on Kauai, Niihau, eastern and western Oahu, eastern Molokai, and probably also the Kohala section of Hawaii, down faulting has been one of the principal causes of the change in the initial outlines.

From the information available, it appears that the rate of removal on Kauai is somewhat more rapid than on any other section of Hawaii, and that, in consequence, the existing relief of the dome is most highly differentiated. Furthermore, extensive engulfment has played a significant part in the generation of the present topography. I believe that it is impossible to regard the landscape of Kauai as the oldest in the windward section of the archipelago; that it belongs to the oldest group is unquestioned, but the last principal volcanic period may have closed as soon on eastern and western Oahu, and West Maui. Niihau I regard as an older mountain than Kauai.

Evidence as yet is insufficient to determine the sequence of extinction of the principal vents in the Hawaiian islands. The problem of the low, semi-desert islands needs special consideration. The small amount of fluvial dissection of these mountains is not a factor of age but of deficiency in rainfall over their surfaces. They may be quite as old as or even older than the deeply dissected domes with which they are associated. Certain general relationships of course are evident; Mauna Loa and Kilauea, the still active volcanoes are the youngest of the group; Hualalai, Mauna Kea, and Haleakala constitute the next series. Of these the principal eruptive period last became extinct in Hualalai; the relative antiquity of Mauna Kea and Haleakala may be debated, though the general concensus of opinion makes Haleakala the older. It is certain, however, that minor activity has prevailed longer on Haleakala than on Mauna Kea. The landscapes of both West Maui and eastern Molokai apparently are somewhat younger than those of eastern or western Oahu or of Kauai, although this is yet to be positively established. As to the relations between Kauai and eastern and western Oahu, little can be said. The place of the four low islands, western Molokai, Lanai, Kahoolawe, and Niihau also has yet to be determined. Wentworth (60) believes that "Lanai is younger than any part of Molokai and is older than east Maui or Kahoolawe. Lanai and western Maui may be of nearly simultaneous origin with the probability that if a difference of age exists western Maui is the older."

SUMMARY

Our present knowledge shows that the windward Hawaiian mountains may be divided into four groups: 1. The still active domes—Mauna Loa and Kilauea; 2. the recently extinct domes—Hualalai, Mauna Kea, and Haleakala; 3. the old, deeply eroded domes of medium elevation—Kohala, West Maui, eastern Molokai, eastern and western Oahu, Kauai; 4. the old, low, arid islands—western Molokai, Lanai, Kahoolawe, and Niihau.

The constructional surfaces of the older domes have been largely destroyed either through the action of erosive agents alone or by erosion and down-faulting; on the younger domes, great sections of the constructional surface remain. Faulting has profoundly modified the initial outlines of eastern Molokai, eastern and western Oahu, Kauai, Niihau, and Kohala. Important changes in outline and topography have resulted from changes of level.

Mauna Loa and Kilauea (Group 1) are still active; their relative ages have been disputed, but the opinion now held by Jaggar, Finch, and others is that Mauna Loa is the older. The order of extinction of the principal vents on the recently extinct domes (Group 2), has been: (1) Haleakala; (2) Mauna Kea; (3) Hualalai. Subordinate activity has taken place at minor centers on Hualalai in the last century, on Haleakala in the eighteenth century, but on Mauna Kea there is no record of a historic eruption.

Regarding the time of close of the last principal volcanism on the older domes, there is little positive information. West Oahu, western Molokai, and West Maui each became extinct prior to the eastern members of these doublets; volcanic activity probably ceased on Niihau before it did on Kauai; Kohala evidently is the oldest dome on Hawaii.

The degree of topographic modification of the older group of domes ranks as follows: 1, Kauai and East Oahu; 2, West Oahu; 3, eastern Molokai; 4, West Maui; 5, Kohala.

The place of the low, arid mountains in the sequence is much more difficult to determine, as little is known of the rates of erosion on them as compared with the higher, older domes. Niihau I believe to be older than Kauai; west Molokai became extinct before east Molokai; Lanai and Kahoolawe both are older than East Maui; Lanai and Kahoolawe probably are of about the same age as West Maui. Beyond these unsatisfactory statements it is unsafe to venture at present.

The principal volcanism apparently ceased on Haleakala, Mauna Kea, and Hualalai in the Pleistocene, though Hualalai may have seen activity at its principal center in Recent times. The older, higher domes doubtless became extinct in the Pliocene and early Pleistocene. The dating of these events must be made from studies of the rates of erosion on the various domes together with the topographic modifications which have taken place. The loss of mass by engulfment also needs full consideration.

RATES OF EROSION IN HAWAII

Erosion by Streams

In Table 14 is given the number of permanent and intermittent streams on the windward and leeward sides of the various Hawaiian domes. The extent to which tributary streams have been developed is also indicated.

MARINE EROSION

The amount of wave erosion about the shores of the Hawaiian domes is an important factor in determining the relative age of the various landscapes. In general, similarly exposed coasts have been abraded at about the same rate, though there is the possibility that, during the Glacial epochs, the speed of removal may have been different than under present climatic conditions. This would affect comparisons of the rates of marine abrasion about islands which

TABLE 14. PERMANENCY OF STREAMS ON THE HAWAIIAN DOMES.

	WINDWARD		LEEWARD		
	Permanent	Intermittent	Permanent	Intermittent	Tributaries
Niihau	....	3	....	21	Few
Kauai	59	2	14	60	Many on large windward and leeward streams; few on most leeward streams.
West Oahu	....	22	....	12	Few
East Oahu	25	7	31	28	Few
West Maui	7	60	3	50	Many on windward and large leeward streams; few on small windward and most leeward streams; some leeward streams permanent in upper courses.
East Maui[a]	....	....	....	....	
West Molokai	....	11	....	27	Few
East Molokai	12	19	2	64	Many in upper courses of windward streams; few on small windward and all leeward streams.
Lanai	....	27	....	16	Few
Kahoolawe[b]	....	....	....	....	
Hawaii					
Kohala	49	6	....	23	Many in upper courses of large windward streams; few in small windward and in all leeward streams.
Mauna Kea	116	20	[b]	....	
Mauna Loa[c]	....	....	....	....	
Hualalai[c]	....	....	....	....	
Kilauea[c]	....	....	....	....	

[a] Permanence of flow of 71 streams not indicated; [b] no record; [c] record incomplete.

had become extinct during the late Tertiary or early Pleistocene and those which became extinct in the late Pleistocene or Recent periods. The rate of removal about the windward shores of Lanai and Kahoolawe probably is slower than that along the windward shores of the other domes, because of the protection afforded these islands by the great bulk of Maui to the east. The strong cliffing of the leeward coasts of Kauai, Niihau, Lanai, and Kahoolowe introduces a special problem. This cliffing may have taken place when the climatic belts were differently arranged than at present. During the Glacial epochs, it is quite possible that the belt of the stormy westerlies may have been shifted as far south at the latitude of Hawaii. Under such conditions, the chief wave action would have been along the leeward rather than along the windward coast as at present. Faulting has played an important part in the development of coastal topography in Hawaii. The eastern coast of Niihau, the northwestern cliffs of Kauai, the magnificent northern face of east Molokai, the northeastern cliffs of Kohala, and part of the southwestern coast of Mauna Loa and Kilauea are fault scarps or fault line scarps. Various sections of the Lanai and Mauna Loa coasts are true fault scarps; the other cliffs probably have been driven back for some distance from the zone of dislocation by the powerful wave attack at their base.

EFFECTS OF EROSION

The climatic and hydrographic data indicate a more rapid rate of erosion over most of Kauai than over any other of the older mountains in the archipelago; therefore the great amount of fluvial removal has not necessarily taken a longer time, but has proceeded at a more rapid rate. Also, a great section of eastern Kauai has been partially destroyed by engulfment. Comparing the series of events recorded in the development of the present landscapes of Kauai and of eastern and western Oahu, it appears that the last principal volcanism ceased first on the Oahu domes. The rate of erosion over eastern Oahu is slower than over Kauai because of the somewhat lighter rainfall and because of the much smaller area of its principal watershed. Western Oahu had been deeply eroded by streams and waves before the growth of the eastern volcano, but, as this mountain was built, it formed a great barrier in the path of the Trade Winds, and gradually lessened the precipitation over western Oahu. The streams became intermittent, and the rate of fluvial removal became relatively slow. Erosion of the eastern mountain has been rapid, and its surface is now more highly dissected than that of its older neighbor. Niihau I believe to be older than Kauai, though the difference in the ages of their landscapes probably is not very great. Morphologic comparisons are difficult because of the vastly different rates of fluvial removal over the two islands. So far as marine removal is concerned, the emerged, wave-cut platforms about the two islands are of about the same width, but that on Kauai is on the windward side, whereas the Niihau platform is on the leeward side. As the average rate of wave erosion probably has been greater on the present windward side of both islands, it would appear that the platform of Niihau has taken a longer time to be developed. However, the Niihau platform did not emerge until after that of Kauai. Whether these opposing factors can be balanced I think is doubtful.

Regarding eastern Molokai, West Maui, and Kohala, I believe that present knowledge of their climates, drainage systems, and morphologic changes indicate that the landscape of West Maui is the oldest, that of eastern Molokai the next, and that of Kohala the youngest. All of these domes became extinct after Kauai, Niihau, and the two volcanoes of Oahu.

The problem of the order of extinction of the various domes merits much further study. The chief points to be borne in mind are the relation between physiographic stage and rate of erosion and the extent of down faulting. In Hawaii the rate of erosion is largely a factor of climate.

NIIHAU

EXTENT OF INVESTIGATIONS

Little geological work had been done on Niihau prior to my visit in 1921. Hitchcock (40) and also Bryan (5) make brief mention of the island and Powers (53), who spent a few hours on Niihau, collected a number of samples of the lavas, some of which have been analyzed by Washington and Keyes (59).

Through the kindness of the owner, Mr. Aubrey Robinson, I was able to spend eight days on Niihau and to visit most parts of it. The time was sufficient for examining the salient features of the geology and morphology and for collecting systematic suites of the igneous rocks, sedimentary rocks, weathered rocks, and soils. At the time of my visit the topographic map of the United States Geological Survey (published in 1929) was not available, hence it was necessary to base the geologic studies on such sketch maps as could be obtained.

GEOGRAPHY

Semidesert Niihau is a roughly elliptical island, lying about 17 miles southwest of Kauai from which its steeply cliffed eastern coast is usually visible. It has an area of 97 square miles, a north-south length of 18 miles, and a maximum width of 5 miles. A small, dissected and strongly cliffed lava dome remnant forms a portion of the eastern side of the island; bordering this on the north, west, and south is a low plain which stands from 50 to 100 feet above sea level. The highest point on the dome remnant is 1,281 feet above sea level.

The only rainfall records are kept at Nonopapa, where an annual maximum of 35 inches has been measured. The average annual precipitation at this station and over the whole island is considerably less. The rains are widely separated in point of time, and months or even more than a year may pass with little or no rain on any part of the island. The low elevation is responsible for the aridity of the climate; only occasionally do cloud caps form about the highland, a condition strikingly contrasted with that on neighboring Kauai. There are no permanent streams; the valleys are filled now and then with torrential floods which may flow for a few days, but for months at a time no water flows through them. The contrast in the climate of Kauai and Niihau naturally is reflected in epigene profiles of the two islands; ocean waves have been a much more effective erosive agent on Niihau than have streams.

Because of its dry climate, Niihau is unforested, and the plant cover in most places is more or less broken. A few scattered native trees and shrubs, such as the milo, the wiliwili, the hau, and the *mao* grow in the highland valleys and on the lowlands at or near the valley mouths. Of the exogenous plants, the algaroba (*Prosopis juliflora*) is most important, because it grows rapidly under arid or semiarid conditions; the wood makes excellent fuel, and the numerous seed pods afford food for cattle and sheep. The principal vegetation is a grass cover, continuous or broken, depending upon the soil in which it grows. On the sand hills along the coast, the *kolakola* and *Convolvulus* are the principal species.

Most of the fresh water on the island is collected from the scanty rainfall. Borings into the rocks of the lowlands have yielded small quantities of fresh water, but the flow soon becomes more or less saline because of the penetration of ocean water through the cavernous and porous rocks, once the head of fresh water had been drawn off. A few brackish springs in the highland provide additional water. On account of insufficient water supply, agriculture is impossible in spite of the evident richness of the soil.

The island is used as a stock ranch. When the scanty rains are not interrupted by too long periods of drought, there is sufficient pasturage for 20,000 to 30,000 head of sheep and more than 1,000 cattle and horses. During the not infrequent long dry spells, the problem of supplying the animals with sufficient water is a serious one, and it is usually found necessary to reduce the number of sheep.

The population of Niihau, according to the census of 1920, was 191; the inhabitants, with the exception of the ranch foreman, are Hawaiians, though of course, many are not full blooded. Originally the island had a large Hawaiian population.

A few roads go over the lowland from the principal settlement at Nonopapa to various parts of the island. There are many trails, and all parts of the island are accessible, as there is no dense cover of vegetation at any place.

There are no natural harbors. The chief landing is offshore at Nonopapa; a second one is off Kii on the east coast. Inter-island boats stop at Niihau at irregular intervals, generally to load or unload cattle or sheep. Communication with Kauai is usually carried on by means of Japanese fishing sampans. Niihau is not connected with Kauai or Honolulu by either wireless or cable.

MORPHOLOGY

General Relations

Niihau, like Kauai, is a single lava dome which was originally larger and higher. Down faulting has caused the subsidence below sea level of the eastern portion of the dome and a considerable part of the residual mass has been destroyed by wave action. Judging from the present shape and size

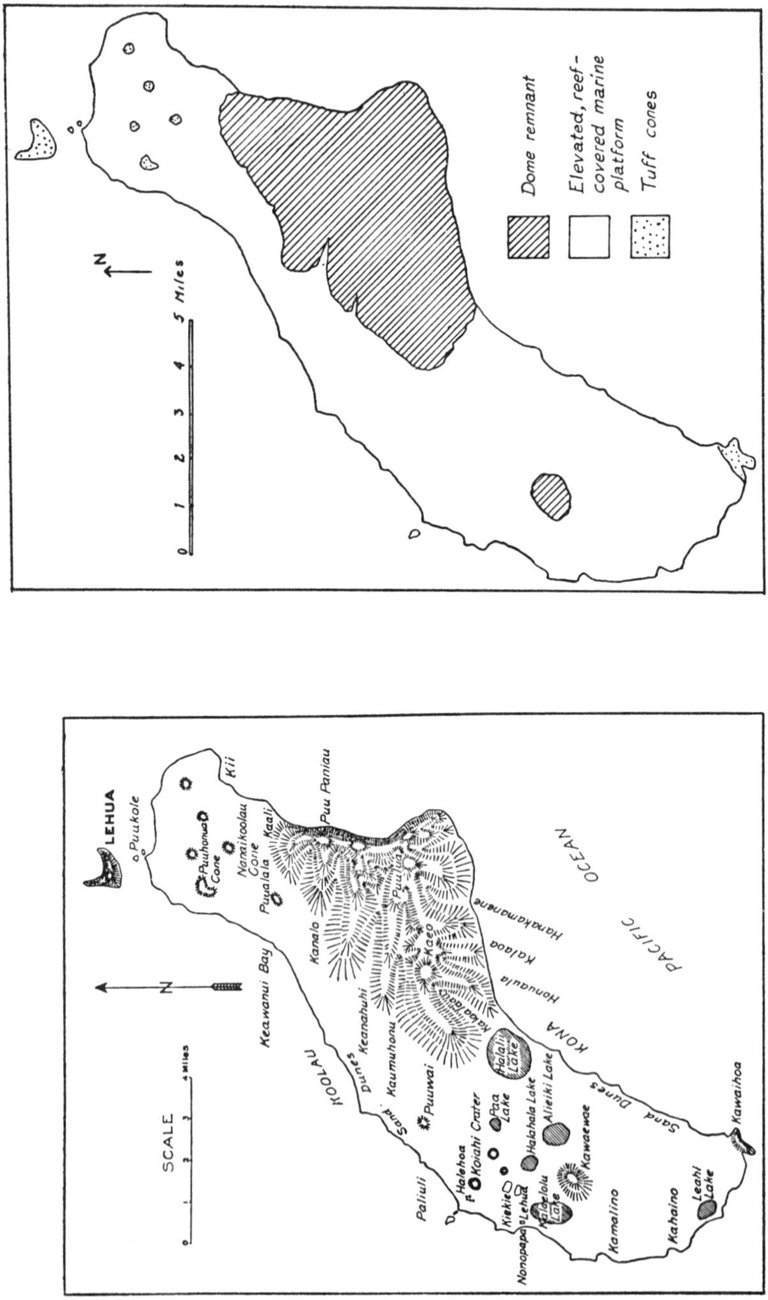

FIGURE 14.—*a*, sketch map of Niihau based on a map published by Powers (Am. Jour. Sci., vol. 50, p. 258, 1920); *b*, physiographic map of Niihau.

of the island, it originally must have been about 18 miles long, 10 miles wide, and more than 3,000 feet high. Its dimensions probably were similar to those of Lanai. Niihau now consists of two sharply differentiated physical divisions (fig. 14): (1) a small, rudely triangular, strongly cliffed and dissected dome remnant composed entirely of westward dipping flows; the northwestern side of the remnant is 9 miles long, the eastern side 4 miles, the southern side 8 miles. Its maximum elevation is 1,281 feet above sea level. (2) Bordering the volcanic mass on the north, west, and south is a low plain whose elevation does not exceed 50 or 100 feet at the base of the highland cliffs (fig. 14, *b*). This plain was formed by the emergence of a wide, wave cut platform partially covered with a thin fringing reef.

Dome Remnant

Because, at a former time, the Niihau dome-remnant was exposed to wave attack on all sides, its margins are strongly cliffed. Marine erosion has been more effective than fluvial erosion, hence wave-cut features are the most conspicuous elements in the relief. The eastern coast, which is a fault line scarp, stands in the path of the northeast trades, and hence is exposed to most active abrasion. The great wall rises to an elevation of about 800 feet at the southern end and to more than 1,000 feet at the northern end. Very few valleys break its face, and the sheerness of the declivity is maintained by the active abrasion at its base (Pl. XI, *A*). The southern cliffs decrease in elevation to about 300 feet at their western extremity. Because this side of the residual is more sheltered, the coast is less steep and more valleys have been eroded into it. On the northwestern side, the cliffs increase in height from about 300 feet at the southern end to more than 1,000 at the northern end. These cliffs have been much more extensively dissected because of the protection long afforded them by the emerged reef-covered platform at their base and because the principal slope of the residual and hence of the drainage is directed toward them.

The scarp forming of the eastern cliffs was produced when more than half of the island was down faulted below sea level (p. 53). Since then, the cliffs have retreated under the powerful wave attack at their base. A submerged platform, about 2 miles wide, indicated by soundings off southeastern Niihau, perhaps measures the amount of wave erosion which has taken place since the engulfment. Within 4 miles of the cliffed coast, depths of 319 to 516 fathoms have been measured. A sharp change in slope thus is suggested at the edge of the platform, and this change I believe represents the position of the rift zone.

Although stream action at the present time is relatively ineffective, the constructional surface has been considerably eroded (Pl. XII, *B*). The time involved in the dissection of the dome remnant is indicated both by the slow rate of fluvial removal and the extent to which waves have attacked the island. As the original elevation of the dome was probably more than 3,000 feet, the precipitation over its surface before the engulfment took place must have been somewhat greater than at present. It may have equalled or exceeded that over Lanai, where a maximum yearly average of 50 inches has been measured. The pre-engulfment streams very likely were much more effective eroding agents than are those of the present, hence it is possible that canyons of considerable size had been eroded into the windward side of the mountain. Whether erosion of the leeward flanks had commenced before the down faulting of eastern Niihau cannot be determined. The existing section is a part of the former leeward portion of the dome. The

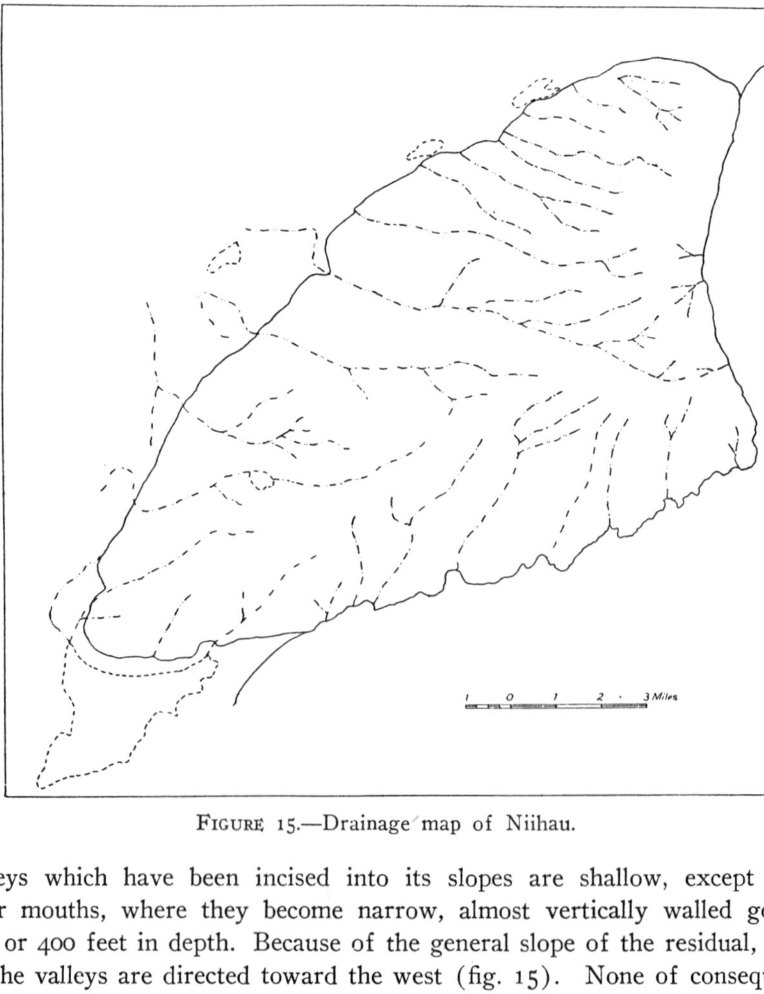

FIGURE 15.—Drainage map of Niihau.

valleys which have been incised into its slopes are shallow, except near their mouths, where they become narrow, almost vertically walled gorges 300 or 400 feet in depth. Because of the general slope of the residual, most of the valleys are directed toward the west (fig. 15). None of consequents has been cut into the eastern cliffs, but a considerable number break the continuity of the southern cliffs. The heads of the valleys lie to the west of the east short cliffs, hence none of them was truncated by the engulfment. Separating the valleys are broad tabular interfluves which are somewhat more dissected in their upstream than in their downstream parts (fig. 16). The ends of the divides have been truncated by wave action before the emergence of the reef-covered platform; because of the arid climate, these former shore cliffs have suffered little dissection.

The largest valley (Keanuahi) and one or two of the smaller ones have been alluviated for a short distance from their mouths, but the depth of the alluvial fill is not known. The presence of these deposits indicates that erosion below the present sea level took place; submergence and alluviation followed the return of sea level to its normal position. This may have happened during the changing levels of the Pleistocene oceans.

Coastal Lowland

The coastal lowland, which borders the dome remnant on the north, west, and south, has been formed by the elevation of a broad, wave-cut platform, covered over a considerable part of its area by a thin, fringing reef (fig. 14, *b*). The width of the plain varies from about 3.5 miles at its northern extremity to 1 mile where it borders the highland on the west and from there increases to over 4 miles in its southern portion. In many places, the reef limestone is absent, and the rather rough basaltic surface is exposed. Also, several, small, cliffed residuals, which were islands in the pre-emergence sea, project above the general level of the plain. The most conspicuous of these is a flat-topped eminence, called Kawaewae, 1 mile southeast of Nonopapa. The top of this residual now stands 290 feet above sea level (fig. 14, *a*).

The surface of the plain is gently rolling, except for certain points which stand considerably above the general level. These points either are pre-emergence islands or tuff cones which have been erupted at various places over the plain. The highest of them, 548 feet, forms the southern tip of the island. In addition, there is a low rampart of sand dunes, 25 to 30 feet high, along considerable stretches of the coast line. At certain localities, these sands have been more or less firmly lithified, and low sea cliffs have been cut into them. No valleys cross the lowland; the occasional flow from the highland gorges is quickly absorbed in the porous soils and rarely, if ever, reaches the sea. Shallow basins of large size, apparently initial depressions in the platform surface or in the reef, are filled with water during the rainy season; in the drier months, the waters partially or completely disappear, and the depressions appear as salt flats, underlain by wet, plastic, lateritic clays, most of which are derived from the weathered basalts of the highland.

In general, the surface of the plain slopes from the base of the dome remnant to the coast line, but at certain localities low cliffs have been cut either into the lavas of the wave cut platform alone or into the overlying reef and lithified sand dunes as well. The most conspicuously cliffed section is in the vicinity of Kiekie. Along the southeastern coast, low cliffs have been cut into the lithified dunes (Pl. XII, *A*). More than two-thirds of the cinder cone (Kawaihoa) at the southern tip of the island has been destroyed by wave erosion.

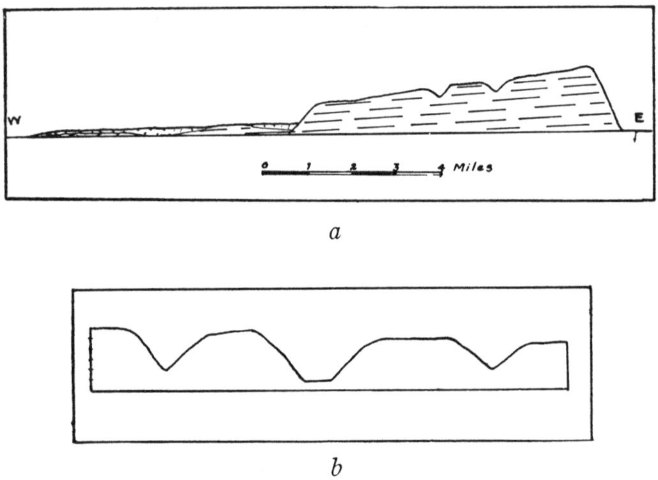

FIGURE 16.—Cross-sections of Niihau. *a*, northeast-southwest cross-section through the island of Niihau. (Vertical scale: 1/10 in. = 200 feet); *b*, valleys and interfluves of the Niihau dome remnant. (Vertical scale: 1 space=100 feet.)

Along the seaward side of Kawaihoa and elsewhere along the coast of Niihau are recently emerged rock terraces standing 4 to 10 feet above sea level. These terraces are similar to those found on most of the other windward Hawaiian islands and are exposed as the result of the Recent 5-meter change of ocean level.

ISOLATED TUFF CONE

Less than a mile from the northern tip of Niihau is the crescentic island, Lehua, a small, partially eroded tuff cone, having a maximum diameter of about 1 mile and an elevation of 702 feet (fig. 14). About 100 degrees of its circumference on the northern side has been destroyed by wave attack. The lower slopes of the cone have been channelled by narrow, shallow gulches, separated by relatively undissected sections of the constructional surface. That a Recent emergence of the cone has taken place is indicated by a narrow, wave-cut bench. As I was not able to visit the island, the amount of change of level was not measured, but it seems to have been considerably more than that which produced the narrow benches about Kauai and Niihau. In the process of elevation, the cone tilted up slightly toward the east.

A second tuff cone, Kaula, lies about 19 miles southwest of Niihau (fig. 12). As described by Palmer (48, pp. 6-10), this cone is built on a platform that finds its origin in the erosion of an original volcano composed of lava. Palmer does not suggest any relationship between the Kaula platform and Niihau, but, that it is a third principal eruptive center of a compound volcano of which Kauai and Niihau are the remaining exposed summit, can scarcely be questioned. Brief accounts of Kaula have also been published by Hitchcock (40), and by Friedlaender (24).

IGNEOUS ROCKS

The Niihau dome remnant is composed of a series of westward dipping lava flows that have been intruded by numerous dikes and sills (fig. 16, *a*).

As on Kauai, the igneous rocks on Niihau are divisible into two series: (1) the main complex, which largely belongs to the principal eruptive period but also includes intrusives of later date; (2) intrusives and extrusives of a Recent volcanic episode. The lavas of the main complex compose the cliffed dome remnant and are exposed at various places over the lowland and along the seacoast. It is possible that the original dome contained a greater variety of petrographic types than are to be found in the existing remnant. The extrusives of the late volcanic episode, tuff cones and lava flows, have been erupted through the reef which overlies the wave-cut platform of west Niihau. Probably the two isolated tuff cones, Lehua and Kaula, also belong to this late episode.

Although the principal rock types are similar megascopically to those of Kauai, the microscope shows that most of them contain more feldspar and less olivine. The four field groups of olivine basalts differentiated on Kauai also are present on Niihau, but their relative importance is different. Coarse-grained plagioclasic chrysophyres (Field group 1) are the dominant lavas; the fine-grained chrysophyres (Field group 2) the chief type of olivine basalt on Kauai, are somewhat less abundant. The lavas of Field group 3 and nephelite basalts, which were found on Kauai, were not obtained from Niihau, though a more thorough search may reveal their presence. A single specimen of nepheline melilite basalt was obtained from a flow belonging to the main complex (34, pp. 534-35).

The flows of the main igneous complex are cut by great number of porphyritic and trappean dikes and sills composed chiefly of olivine basalt and olivine-poor or olivine-free augite andesite. A few diabasic dikes also were found. No coarsely granular plutonic rocks or xenoliths were found.

The lavas of the late episode consist of flows and pyroclastics erupted through the fringing reef on the western side of the island. Some of the dikes and sills cutting the main complex doubtless belong to this series. In the tuff cone (Kawaihoa) fragments of the coral reef are embedded.

Chemical analysis of lavas from Niihau are given in Table 9.

SEDIMENTARY ROCKS

The sedimentary rocks on Niihau are principally on the lowland, but small deposits are present in some of the highland gorges. They include (1) fringing reefs; (2) beach sediments; (3) sand dunes; and (4) alluvial deposits.

1. Much of the elevated wave cut platform is covered by a fringing reef which in few places exceeds 100 feet in thickness. Over considerable sections of the platform, the reef is absent. Most of the reef and the basaltic surface, where exposed, is covered by thin layers of sediments, the greater part of which has been brought down by streams from the highland. Along the coast, the mantle is chiefly of calcareous sand, with which is mixed small quantities of basaltic detritus. So far as could be ascertained, no living reef is present about Niihau. The last volcanic episode occurred after the growth of most or all of the reef. Fragments of reef limestone in some of the tuff cones, and deposits of ejecta overlying the reef duplicate the relations found in Kauai. Reef limestone fragments have been reported also from Kaula.

2. The beach sediments are composed of basaltic and calcareous débris; the dominant constituent varies from place to place.

3. Ramparts of calcareous sand dunes, few of them more than 30 feet in height, are present at many places along the shore, and, at certain localities, moderately thick deposits of sand have been drifted inland close to the base of the dome remnant. The principal areas of lithified dunes is along the southeastern (Kona) coast (Pl. XI, *B*).

The fossils in the following list were obtained from sand dunes near Kii (column 1); from partially lithified dunes near Cape Kawaihoa (column 2); from sand dunes at the south end of the basalt cliffs (column 3); from sand dunes south of Kii (column 4); from a beach on the southeast coast (column 5); and from a beach east of Kawaewae (column 6). The species were identified by Professor Junius Henderson.

	1	2	3	4	5	6	7
Podophera pedifera	—	—	—	—	—	X	—
Echinoid spines	X	—	—	—	—	—	—
Chama sp.	—	—	—	—	X	—	—
Codakia ramulosa Gould	—	—	—	—	X	—	—
Perna sp.	—	—	—	—	X	—	—
Venus reticulatus Linneaus	—	—	X	—	X	—	—
Conus hebraeus Linneaus	—	—	X	—	X	—	—
Conus lividus Hwass	—	—	X	X	X	X	—
Cypra caput-serpentis Linnaeus	—	—	X	—	X	X	X
Cypra isabella Linnaeus	—	—	—	—	X	X	—
Cypra madagascarensis Gmelin	—	—	—	—	X	—	—
Cypra mauritiana Linnaeus	X	—	X	—	—	—	—
Cypra reticulata Martyn	X	—	X	—	X	—	—
Eulima sp.	—	—	—	—	X	—	—
Glyphis granifera Pease	—	—	—	—	X	—	—
Helcioniscus argentatus Sowerby	X	—	X	X	X	—	—
Helcioniscus exaratus Nuttall	X	X	X	X	X	X	X
Hipponyx antiquatus Linnaeus	—	—	—	—	X	—	—
Hipponyx cf. barbatus Sowerby	—	—	—	—	X	—	—
Littorina pintado Wood	—	X	X	—	—	X	—
Nerita picea Recluz	X	X	X	X	X	—	X
Nerita polita Linnaeus	X	—	X	X	—	—	X
Patella sp.	—	X	X	—	X	X	—
Vermetus sp.	—	—	—	—	X	X	—
Purpura aperta Blainville	X	X	X	X	—	—	X
Ricinula ricinus Linnaeus	X	—	X	—	—	X	—
Sistrum tuberculatum Blainville	X	—	—	—	—	—	—
Strombus maculatus Nuttall	X	—	—	—	—	—	—
Terebra strigilata Linnaeus	—	—	X	—	—	—	—
Turbo chrysostomus Linnaeus	X	—	—	X	—	—	X

4. Alluvial Deposits. Of the alluvial deposits, the alluvio-lacustrine sediments on the reef and platform surface are the most worthy of consideration. A number of large, shallow depressions are present in the reef and platform; these in the wet seasons are filled with shallow depths of water. (See p. 95.) Because of the abundant migration of sea water through the porous reef limestone or through the underlying porous and cavernous lavas, the water in the depressions quickly becomes distinctly brackish, and, after the rainy season ends, turns saline upon some evaporation. The depressions are floored with reddish, lateritic clays evidently transported from the highland by wind and by the occasional streams which flow through the valleys. The red clay is abundantly mixed with salt which crystallizes out as the waters disappear; this is specially noticeable during the long dry spells to which the island is rather frequently subject. During the dry periods, the clay hardens and cracks at the surface, but remains wet and plastic below, so that travelling across the flats is dangerous because of the chance of breaking through the crust and becoming mired in the rather deep muds underneath. Cattle and sheep are not infrequently lost in this way.

EVIDENCES OF CHANGES OF LEVEL

The changes of level recorded on Niihau are: the elevation of the wide, wave-cut platform with its thin cover of reef, and the Recent slight emergence of narrow rock terraces.

The elevation of the platform at the base of the dome remnant is about 50 feet. Residuals, which were islands in the pre-emergence sea, rise to a maximum of 290 feet. The vertical movement indicated by the emergence of the platform is 65 to 70 feet. Certain poorly developed terraces cut into the platform indicate that the elevation was interrupted by short periods of still stand. The major change of level recorded on Niihau appears to have been considerably less than on Kauai, where an upward movement of over 500 feet took place. The two movements were not contemporaneous, and the Kauai platform probably was elevated first. No reef is present on the Kauai structure, and into it numerous gorges have been cut. The Niihau bench, on the other hand, has a reef coating, and is virtually undissected by streams. The topographic difference, however, may have resulted from the much greater rainfall over Kauai. Judging from the amount of dissection of the Kauai platform, its emergence must have taken place well back in the Pleistocene at the time of the faulting which broke the Kauai-Niihau volcano in two. The elevation of Niihau took place without tilting, and apparently occurred at a much later date. If the Hawaiian reefs are late Glacial or post-Glacial in age, then Niihau was elevated in late Pleistocene or Recent times.

The Recent 16-foot change of ocean level is recorded on Niihau at many places by rock benches which stand from from 4 to 5 to 8 or 10 feet above mean sea level.

FAULTING

The only positive evidence of faulting on Niihau is that which resulted in the engulfment of the eastern half of the dome. The rift zone lies about 2 miles east of the present eastern shore of the island. Since this dislocation took place, Kauai and Niihau have behaved as more or less distinct units.

THE AGE OF NIIHAU LANDSCAPE

The relative ages of the Kauai and Niihau landscapes are difficult to ascertain. On the doublet island of Oahu, Maui, and Molokai, the western member is the older; if this order of extinction of the last principal eruptive period can be applied to the Kauai-Niihau volcano, Niihau is the older of the two. The third principal eruptive center of the same volcano indicated by the platform underlying Kaula to the southwest of Niihau became extinct at a still earlier period. Niihau seems to occupy an intermediate stage in the postvolcanic history of lava domes between Kauai and the third, submerged center.

The width of the elevated, wave-cut platform on Niihau suggests a long period of erosion, and erosion when the waves were much more active about the leeward shores of the island than they are at the present time. If changed climatic conditions prevailed in the latitudes of the Hawaiian Archipelago during the glacial stages of the Pleistocene, then such features as this platform, the great western cliffs of Kauai and other islands can be more easily accounted for. The last major eruptive cycle on Niihau therefore appears to have closed well back in the Pleistocene or more probably in the Pliocene. Topographic comparisons with Kauai are difficult because of the strongly contrasted climatic conditions over the two islands which has made the topographic evolution very different. On Niihau, marine erosion has been more effective than fluvial, while, on Kauai, the fluvial exceeds marine. Elevated, wave-cut platforms of considerable width are present on western Niihau and eastern Kauai. A deeply submerged platform on the western side of Kauai is indicated by borings, but nothing is known of its width. Kauai has suffered much greater dissections by streams than has Niihau.

BIBLIOGRAPHY

1. BAILEY, C. T. and STEWART, J. E., Surface water supply of Hawaii: U.S. Geol. Survey, Water-Supply Paper 515, 1923.
2. BEALS, E. A., Climatological data, Hawaii Section, Annual Summaries: U.S. Dept. of Agr., 1923, 1924.
3. BLACKWELDER, E., Exfoliation as a phase of rock weathering: Am. Jour. Geol., vol. 33, pp. 793-806, 1925.
4. BRANNER, J. C., Notes on the geology of the Hawaiian islands: Jour. Sci., vol. 16, pp. 301-316, 1903.
5. BRYAN, W. A., The natural history of Hawaii, Honolulu, 1915.
6. CAMPBELL, D. H., Some botanical and environmental aspects of Hawaii: Ecology, vol. 1, pp. 257-269, 1920.
7. CROSS, W., Lavas of Hawaii and their relations: U.S. Geol. Survey, Prof. Paper 88, 1915.
8. DAINGERFIELD, L. H., Summary of climatological data for the United States by Sections, Hawaii Section: U.S. Dept. of Agr., pp. 1-52, 1918.
9. DAINGERFIELD, L. H. Climatological data, Hawaii Section, Annual Summaries: U.S. Dept. of Agr., 1915-1921.
10. DAINGERFIELD, L. H., Kona storms: Monthly Weather Review, pp. 327-329, 1921.
11. DALY, R. A., Igneous rocks and their origin, New York, 1913.
12. DALY, R. A., The geology of American Samoa: Carnegie Inst. Washington, Yearbook 18, pp. 192-195, 1919.
13. DALY, R. A., A general sinking of sea level in recent times: Nat. Acad. Sci., Proc. vol. 6, pp. 246-250, 1918.
14. DALY, R. A., The geology of Ascension and St. Helena islands: Geol. Mag., vol. 59, pp. 146-156, 1922.
15. DALY, R. A., A recent world-wide sinking of ocean level: Geol. Mag., vol. 57, pp. 246-257, 1920.
16. DALY, R. A., The geology of Ascension Island: Am. Acad. Arts and Sci., Proc., vol. 60, pp. 4-80, 1925.
17. DALY, R. A., The geology of St. Helena Island: Am. Acad. Arts and Sci., Proc., vol. 62, pp. 31-92, 1927.
18. DANA, E. S., Contributions to the petrography of the Sandwich Islands: Am. Jour. Sci., vol. 37, pp. 441-467, 1889. Also in Characteristics of volcanoes, New York, pp. 318-354, 1890.
19. DANA, J. D., United States Exploring Expedition, 1838-1842, vol. 10, Geology, pp. 262-279, 1849.
20. DANA, J. D., A dissected volcanic mountain, some of its relations: Am. Jour. Sci., vol. 32, pp. 247-255, 1886.
32. HINDS, N. E. A., The geology of Kauai, Hawaiian islands: Geol. Soc. Am., Bull., vol. 33, p. 125, 1921.
33. HINDS, N. E. A., The geology of Kauai and Niihau, Hawaiian islands: Zeitschr. fur Vulkanologie, vol. 12, 1929.
34. HINDS, N. E. A., Melilite and nephelite basalt in Hawaii: Jour. Geol., vol. 33, pp. 526-539, 1925.
35. HINDS, N. E. A., Amphitheatre valley heads: Jour. Geol., vol. 33, pp. 816-818, 1925.
36. HINDS, N. E. A., Maui and the Maui group, Hawaii: Geog. Soc., Philadelphia, Bull., vol. 23, pp. 147-165, 1925.
37. HINDS, N. E. A., Weathering of Hawaiian lavas: Am. Jour. Sci., 5th ser., vol. 17, pp. 297-320, 1929.
38. HITCHCOCK, C. H., The geology of Oahu: Geol. Soc., Am., Bull., vol. 11, pp. 15-60, 1900.
39. HITCHCOCK, C. H., The geology of Diamond Head, Oahu: Geol. Soc., Am., Bull., vol. 17, pp. 469-484, 1906.
40. HITCHCOCK, C. H., Hawaii and its volcanoes, Honolulu, 1909.
41. KELLEY, W. P., Rice soils of Hawaii: Hawaii Agr. Exper. Station, Bull. 31, 1914.
21. DANA, J. D., Characteristics of volcanoes, New York, 1890.
22. DAVIS, W. M., The marginal belts of the coral seas: Am. Jour. Sci., vol. 6, pp. 181-195, 1923.

23. DUTTON, C. E., Hawaiian volcanoes, U.S. Geol. Survey, Fourth Ann. Rept., pp. 75-219, 1884.
24. FRIEDLAENDER, I., Die Insel Kaula in der Hawaiigruppe: Zeit. fur Vulkanologie, vol. 7, pp. 107-108, 1923.
25. FREEMAN, O. W., The origin of Swimming Woman Canyon, Big Snowy Mountains, Montana: Jour. Geol., vol. 33, pp. 75-79, 1925.
26. GROVER, N. C., Surface water supply of Hawaii: U.S. Geol. Survey, Water-Supply Paper 445, 1927.
27. GROVER, N. C., Surface water supply of Hawaii: U.S. Geol. Survey, Water-Supply Paper 465, 1918.
28. GROVER, N. C., and LARRISON, C. K., Surface water supply of Hawaii: U.S. Geol. Survey, Water-Supply Paper 430, 1917.
29. HAMRICK, A. M., Cumulus clouds in Hawaii: Monthly Weather Review, pp. 415-417, 1918.
30. HILLEBRAND, W., Flora of the Hawaiian islands, vol. 1, London, 1888.
31. HENRY, A. J., Increase of precipitation with altitude: Monthly Weather Review, pp. 33-41, 1919.
42. KELLEY, W. P., and MCGEORGE, W. T., The soils of the Hawaiian islands: Hawaii Agr. Exper. Station, Bull. 40, 1915.
43. LINDGREN, W., Water resources of Molokai, Hawaiian islands: U.S. Geol. Survey, Water-Supply Paper 77, 1903.
44. LYONS, A. B., Chemical composition of Hawaiian soils and the lavas from which they have been derived: Am. Jour. Sci., vol. 2, pp. 421-429, 1896.
45. MARTIN, W. F., and PIERCE, C. H., Water resources of Hawaii: U.S. Geol. Survey, Water-Supply Paper 318, 1913.
46. MCGEORGE, W. T., Composition of Hawaiian soil particles: Hawaii Agr. Exp. Station, Bull. 42, 1917.
47. MÖHLE, R., Beitrag zur petrographic der Sandwich und Samoa iseln: Neues Jahrb., Beilage vol. 15, pp. 66-104, 1902.
48. PALMER, H. S., Geology of Kaula, Nihoa, Necker, and Gardner and French Frigates Shoal: B. P. Bishop Museum, Bull. 35, 1927.
49. PILSBRY, H. A., Mid-Pacific land-snail faunas: Manual of conchology, vol. 21, Philadelphia, 1911.
50. PILSBRY, H. A., Mid-Pacific land-snail faunas: Nat. Acad. Sci., Proc., vol. 2, pp. 429-433, 1916.
51. POWERS, S., A lava tube at Kilauea, Tulsa, Oklahoma, 1914.
52. POWERS, S., Tectonic lines in the Hawaiian islands: Geol. Soc., Am., Bull., vol. 26, 501-514, 1917.
53. POWERS, S., Notes on Hawaiian petrology: Am. Jour. Sci., vol. 50, pp. 256-280, 1920.
54. SETCHELL, W. A., Personal communication.
55. STEWART, J. E., Surface water supply of Hawaii: U.S. Geol. Survey, Water-Supply Paper 516, 1924.
56. STEWART, J. E., Surface water supply of Hawaii: U.S. Geol. Survey, Water-Supply Paper 535, 1924.
57. STONE, J. B., The products and structure of Kilauea: B. P. Bishop Mus., Bull. 133, 1926.
58. WASHINGTON, H. S., Petrology of the Hawaiian islands, I: Am. Jour. Sci., vol. 5, pp. 465-502, 1923; II, vol. 6, pp. 100-126, 1923; III, pp. 338-367, 1923; IV, pp. 409-423, 1923; V, vol. 12, pp. 336-352, 1926.
59. WASHINGTON, H. S., and KEYES, M. G., Petrology of the Hawaiian islands: VI, The Leeward islands: Am. Jour. Sci., vol. 12, pp. 199-220, 1928.
60. WENTWORTH, C. K., The geology of Lanai: B. P. Bishop Mus., Bull. 24, 1925.
61. WENTWORTH, C. K., The pyroclastic geology of Oahu: B. P. Bishop Mus., Bull. 30, 1926.
62. WENTWORTH, C. K., Estimates of marine and fluvial erosion in Hawaii: Jour. Geol., vol. 35, pp. 117-135, 1927.
63. WENTWORTH, C. K., and PALMER, H. S., Eustatic bench in the islands of the North Pacific: Geol. Soc., Am., Bull., vol. 34, pp. 521-544, 1925.
64. VON WOLFF, F., Der Vulkanismus, Stuttgart, 1913.

A

B

A, ISLAND OF LANAI (PHOTOGRAPH BY R. J. BAKER); *B*, WAIMEA CANYON, SOUTHERN KAUAI.

A

B

A, HANGING VALLEY ALONG THE CLIFFED COAST OF NORTHWESTERN KAUAI; *B*, DISCORDANT JUNCTION OF TRIBUTARY AND THE MAIN STREAM, HANAPEPE CANYON.

A

B

C

A, LOWER PORTION OF OLOKELE CANYON; *B*, VIEW OF THE DISSECTED HIGHLAND OF SOUTHERN KAUAI (PHOTOGRAPH BY O. S. EMERSON); *C*, PANORAMIC VIEW OF KOLALAU CANYON OF NORTHWESTERN KAUAI (PHOTOGRAPH BY O. S. EMERSON).

A

B

C

A, TERMINUS OF A SHARPLY SERRATED RIDGE IN THE DISSECTED HIGHLAND OF NORTHERN KAUAI; *B*, DISSECTED, WAVE-FACETTED SPURS ALONG THE INNER MARGIN OF THE EMERGED MARINE PLATFORM, NORTHERN KAUAI (PHOTOGRAPH BY R. J. BAKER); *C*, YOUNG GORGE CUT BY SOUTH FORK OF WAILUA RIVER CROSSING THE EMERGED MARINE PLATFORM OF EASTERN KAUAI.

A

B

C

A, THE ALLUVIATED FLOOR OF HANALEI RIVER; *B*. HANAMAULU VALLEY. AN ALLUVIATED GORGE ERODED INTO THE MARINE PLATFORM OF EASTERN KAUAI; *C*, TUFF CONE (KILOHANO) ERUPTED ON THE MARINE PLATFORM OF EASTERN KAUAI (PHOTOGRAPHS BY R. J. BAKER).

A

B

C

A, SHOREWARD SIDE OF THE DISSECTED TUFF CONE AT KILAUEA BAY; *B*, COASTAL MARGIN OF THE CONSTRUCTIONAL, HINTER-REEF PLAIN OF NORTHERN KAUAI; *C*, SEAWARD SLOPING, EMERGED MARINE PLATFORM NEAR MOUTH OF KALIHIWAI STREAM (PHOTOGRAPH BY SENDA).

A

B

C

A, RAMPART OF SAND DUNES ALONG THE OUTER MARGIN OF THE HINTER-REEF PLAIN, NORTHERN KAUAI; *B,* FRINGING REEF, NORTHERN KAUAI; SEA CLIFFS ERODED INTO THE EMERGED, WAVE-CUT, PLATFORM OF NORTHERN KAUAI (PHOTOGRAPH BY SENDA).

A

B

A, DISSECTED, RECENTLY NIPPED CLIFFS FORMING THE NORTHWESTERN (NAPALI) COAST OF KAUAI (PHOTOGRAPH BY R. J. BAKER); *B*, WAVE-CUT TUNNEL THROUGH A NARROW HEADLAND OF THE NORTHWESTERN (NAPALI) COAST OF KAUAI.

A

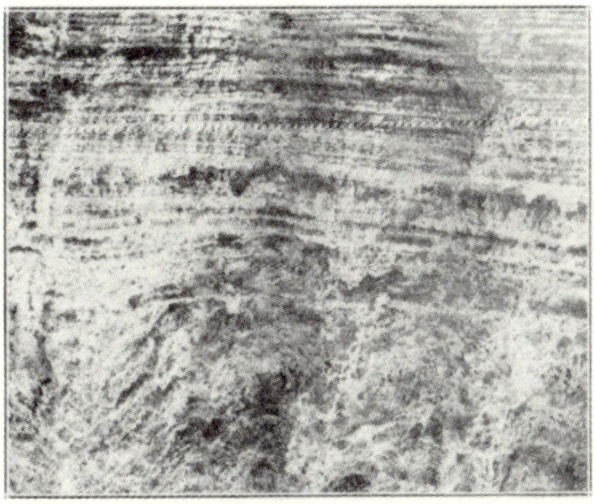

B

C

A, INNER MARGIN OF THE CONSTRUCTIONAL HINTER-REEF PLAIN OF NORTHERN KAUAI; *B*, *C*, STRUCTURES IN LITHIFIED SAND DUNES, SOUTHERN COAST OF KAUAI.

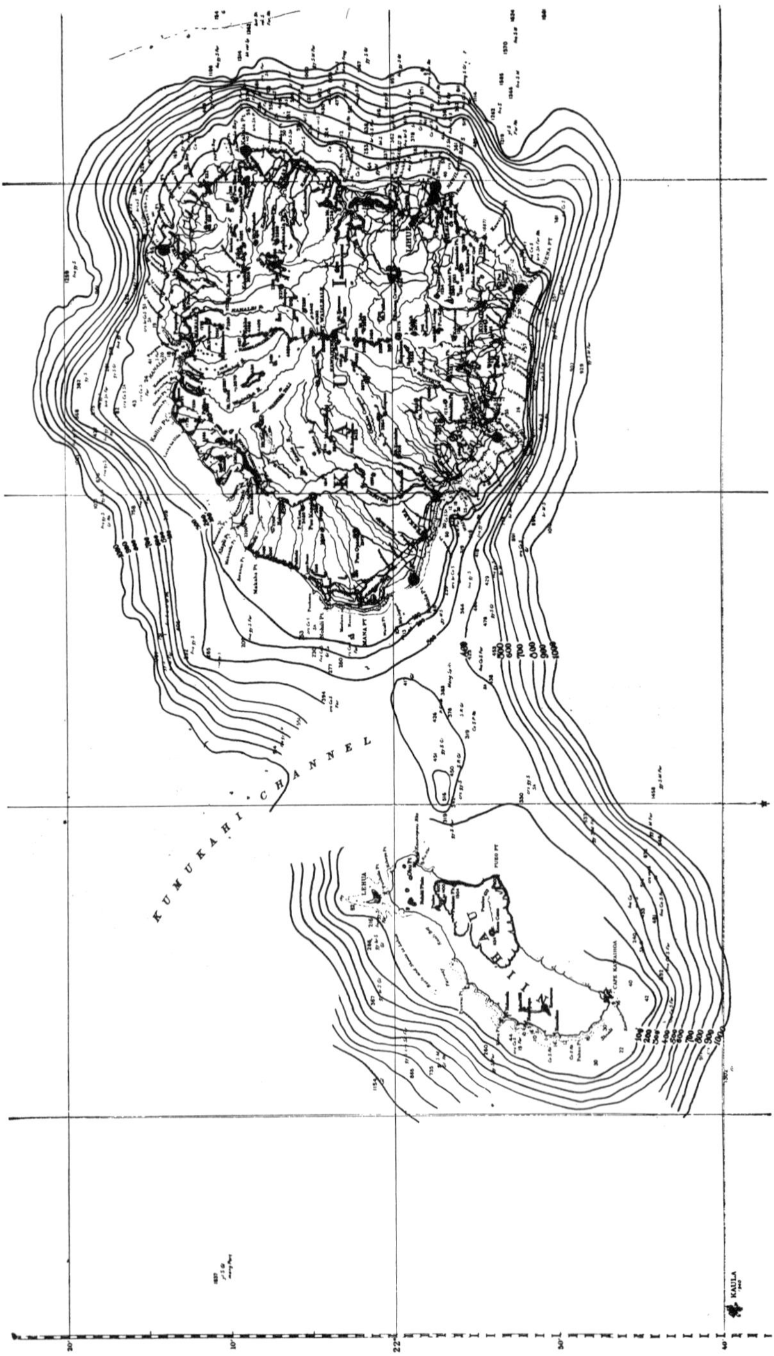

KAUAI-NIIHAU VOLCANO, SUBMARINE CONTOURS 100-FATHOMS. DATA FROM COAST AND GEODETIC SURVEY.

A

B

A, FAULT LINE, SCARP FORMING THE EASTERN FACE OF THE NIIHAU LAVA DOME REMNANT; *B*, DISSECTED TUFF CONE (KAWAIHOA), SOUTHERN END OF NIIHAU.

A

B

A, THE EMERGED, REEF-COVERED PLAIN OF SOUTHERN NIIHAU; *B*, GENERAL VIEW OF THE NIIHAU DOME REMNANT.